Science Notebook

Course 2

Consultant
Douglas Fisher, Ph.D

About Douglas Fisher

Douglas Fisher, Ph.D., is a professor at the Department of Teacher Education at San Diego State University and is a teacher at Health Sciences High & Middle College. He is the recipient of an International Reading Association Celebrate Literacy Award, as well as a Christa McAuliffe Award for Excellence in Teacher Education. He has published numerous articles on reading and literacy, differentiated instruction, English-Language learners, and curriculum design, as well as books such as *Improving Adolescent Literacy: Strategies at Work, Checking for Understanding: Formative Assessment Tools for your Classroom, Productive Group Work,* and *Enhancing RtI*. He has taught a variety of courses in SDSU's teacher credentialing program as well as graduate-level courses on English-Language development and literacy. He also teaches classes in English, writing, and literacy developments to secondary school students.

The McGraw·Hill Companies

Send all inquiries to:
McGraw-Hill Education
8787 Orion Place
Columbus, OH 43240-4027

ISBN: 978-0-07-889431-2
MHID: 0-07-889431-X

Printed in the United States of America.

6 7 8 9 10 RHR 16 15 14

Table of Contents

Using Your Science Notebook

This note-taking guide is designed to help you succeed in learning science content. Each chapter includes:

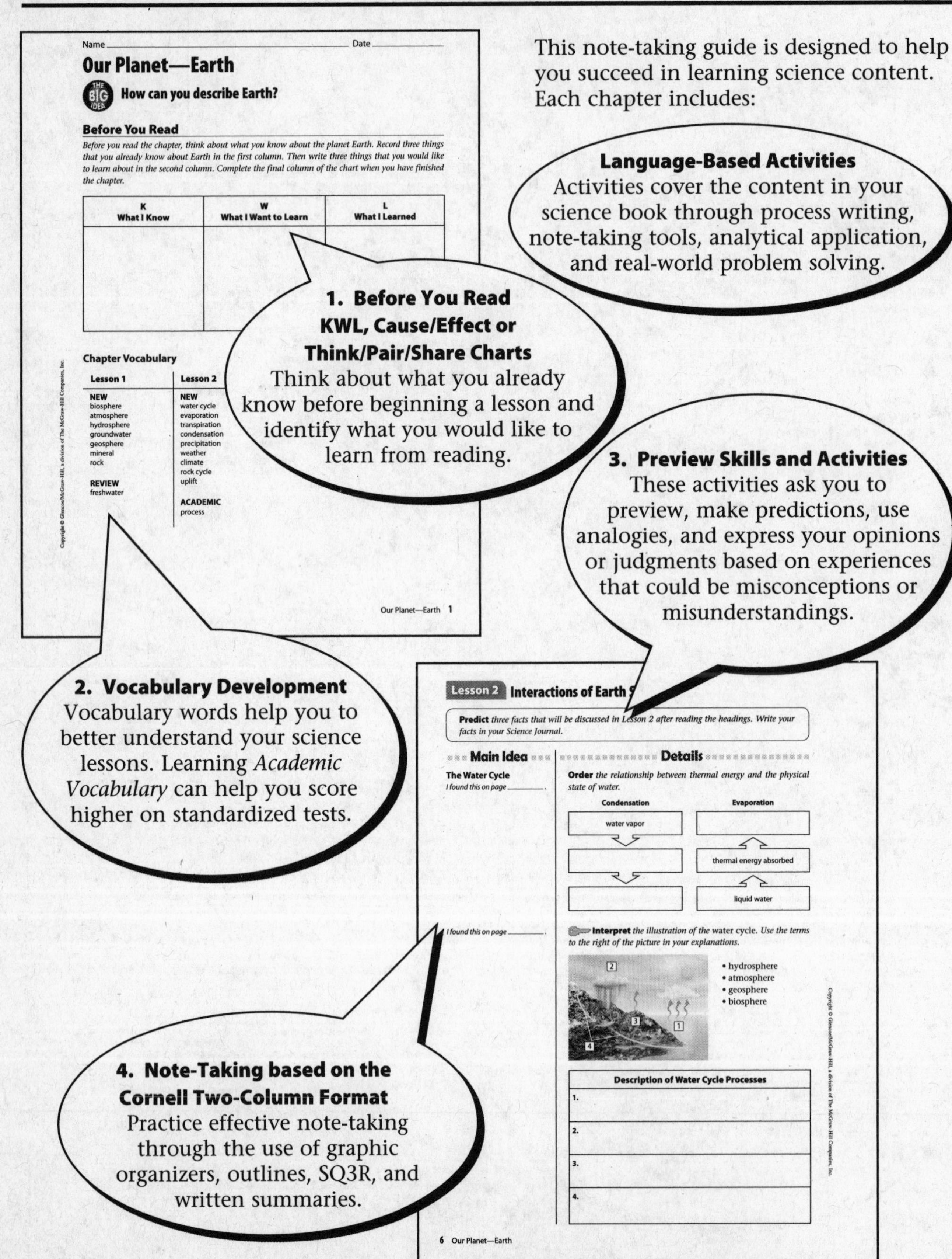

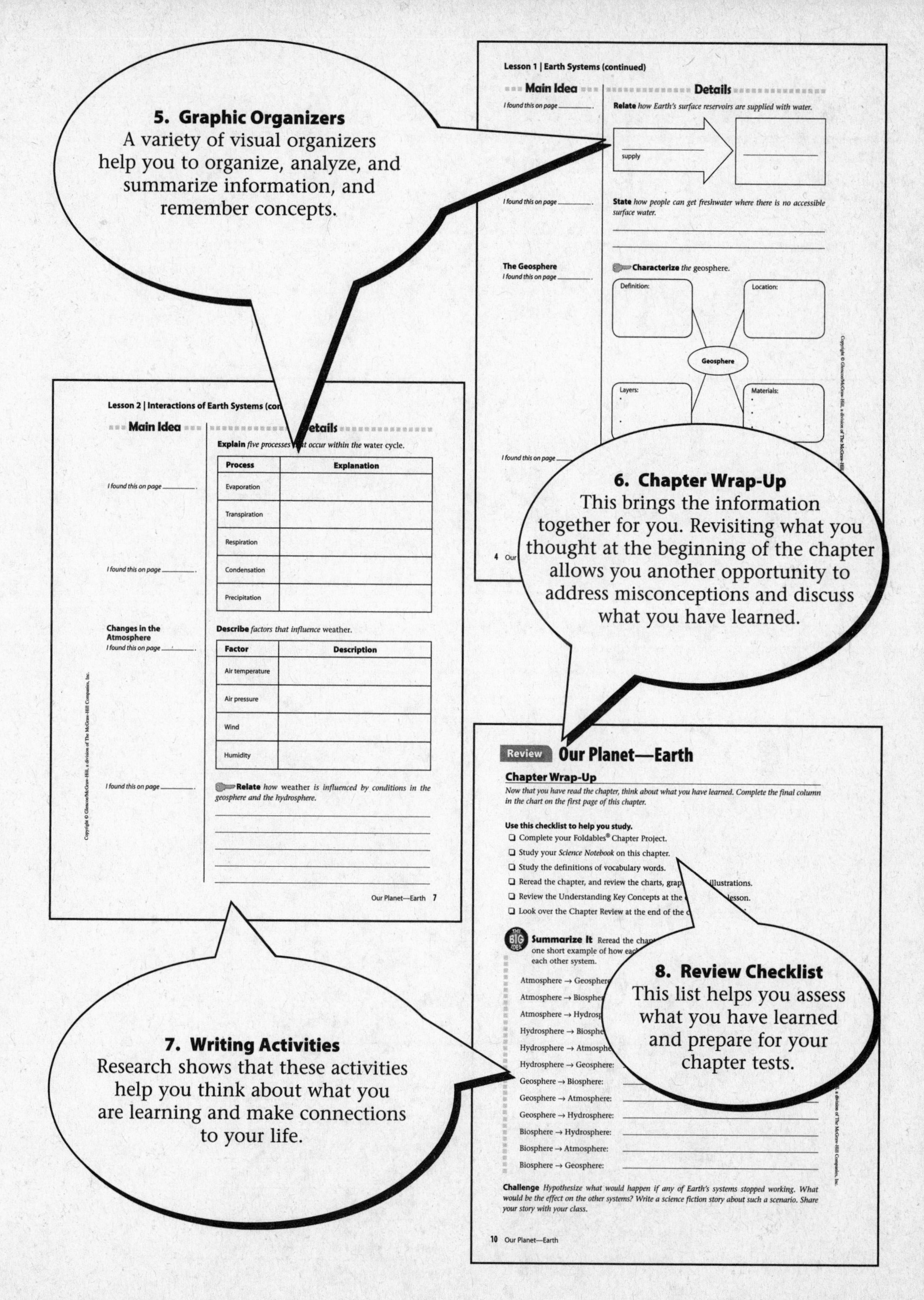
Lesson 1 | Earth Systems (continued)
Main Idea
Details
I found this on page
Relate how Earth's surface reservoirs are supplied with water.
supply
State how people can get freshwater where there is no accessible surface water.
The Geosphere
Characterize the geosphere.
Definition:
Location:
Geosphere
Layers:
Materials:
5. Graphic Organizers
A variety of visual organizers help you to organize, analyze, and summarize information, and remember concepts.
Lesson 2 | Interactions of Earth Systems
Explain five processes that occur within the water cycle.
Process
Explanation
Evaporation
Transpiration
Respiration
Condensation
Precipitation
Changes in the Atmosphere
Describe factors that influence weather.
Factor
Description
Air temperature
Air pressure
Wind
Humidity
Relate how weather is influenced by conditions in the geosphere and the hydrosphere.
Our Planet—Earth 7
6. Chapter Wrap-Up
This brings the information together for you. Revisiting what you thought at the beginning of the chapter allows you another opportunity to address misconceptions and discuss what you have learned.
Review
Our Planet—Earth
Chapter Wrap-Up
Now that you have read the chapter, think about what you have learned. Complete the final column in the chart on the first page of this chapter.
Use this checklist to help you study.
Complete your Foldables® Chapter Project.
Study your Science Notebook on this chapter.
Study the definitions of vocabulary words.
Summarize It
7. Writing Activities
Research shows that these activities help you think about what you are learning and make connections to your life.
8. Review Checklist
This list helps you assess what you have learned and prepare for your chapter tests.
Hydrosphere → Geosphere:
Geosphere → Biosphere:
Geosphere → Atmosphere:
Geosphere → Hydrosphere:
Biosphere → Hydrosphere:
Biosphere → Atmosphere:
Biosphere → Geosphere:
Challenge Hypothesize what would happen if any of Earth's systems stopped working. What would be the effect on the other systems? Write a science fiction story about such a scenario. Share your story with your class.
10 Our Planet—Earth

Name ______________________________ Date ____________

Scientific Explanations

How can science provide answers to your questions about the world around you?

Before You Read

Before you read the chapter, think about what you know about how science provides answers to questions about the world. Record your ideas in the first column. Pair with a partner, and discuss his or her thoughts. Write those ideas in the second column. Then record what you both would like to share with the class in the third column.

Think	Pair	Share

Chapter Vocabulary

Lesson 1	Lesson 2	Lesson 3
NEW observation hypothesis prediction inference technology scientific theory scientific law critical thinking **ACADEMIC** ethics	**NEW** description explanation International System of Units (SI) accuracy precision significant digits	**NEW** variable dependent variable independent variable constants

Lesson 1 Understanding Science

Predict *three facts that will be discussed in Lesson 1 after reading the headings. Record your predictions in your Science Journal.*

Main Idea	Details
What is science? *I found this on page* ______.	**Identify** *three behaviors that scientists might use in exploring questions and in solving problems.* **1.** ______ **3.** ______ **2.** ______
Branches of Science *I found this on page* ______.	**Differentiate** *3 main branches of science. Describe what scientists study in each area.*

Main Idea	Details
Scientific Inquiry	**Define** *terms applied to scientific inquiry.*
I found this on page ______.	
I found this on page ______.	

Observation	Hypothesis
Prediction	**Inference**

I found this on page ______.

Practice *stating a research* hypothesis. *Write a research* hypothesis *that might form the basis of an investigation.*

______

______

______

Main Idea	Details

Results of Scientific Inquiry

I found this on page ________.

Categorize *outcomes of scientific inquiry. Give an example of each type of result.*

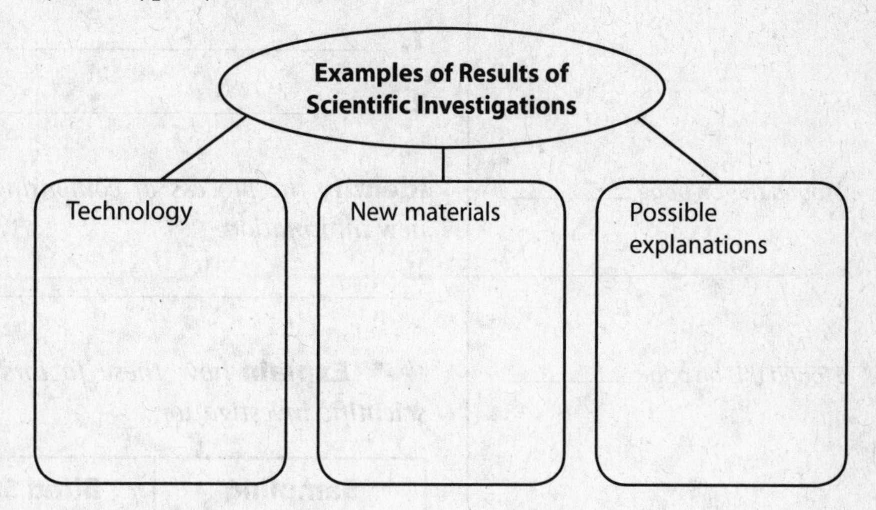

Scientific Theory and Scientific Laws

I found this on page ________.

Contrast *a* scientific theory *with a* scientific law.

Scientific Theory	Scientific Law
Description:	Description:
Example:	Example:

I found this on page ________.

Relate *two ways in which a* scientific theory *and a* scientific law *are similar.*

Both...
- will be rejected if
- are based on

Main Idea | Details

I found this on page ______.

Assess *two situations in which it is important to question scientific issues in the media.*

1. ______
2. ______

I found this on page ______.

Identify *the process of comparing what you already know with new information.*

______

I found this on page ______.

Explain *how these factors can help minimize bias in a scientific investigation.*

Sampling	Blind Study	Repetition

I found this on page ______.

Point out *the importance of safety and ethics when conducting scientific investigations.*

Safety	Ethics

Analyze It Suppose you see two news stories on TV about a scientific topic, but the experts say different things about the data and what it means. How do you decide which is correct?

______

______

______

______

Lesson 2 Measurement and Scientific Tools

Scan *Lesson 2. Read the lesson titles and bold words. Look at the pictures. Identify three facts you discovered about measurement and scientific tools. Record your facts in your Science Journal.*

Main Idea	Details

Description and Explanation

I found this on page ________.

Relate *the terms* description *and* explanation *to observations.*

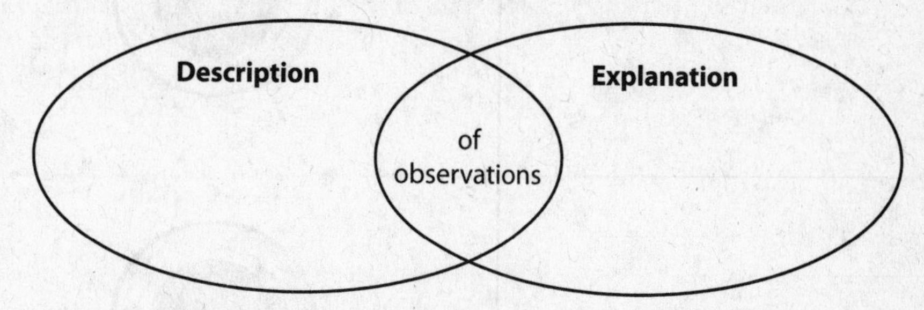

The International System of Units

I found this on page ________.

Interpret *the mathematical meaning of prefixes used in the* International System of Units (SI).

Prefix	Meaning	Prefix	Meaning
Mega		Micro	
Kilo		Milli	
Hecto		Centi	
Deka		Deci	

I found this on page ________.

Identify *the* SI *units for different measurements.*

Quantity Measured	Unit
Length	
Mass	
Time	
Electric current	
Temperature	
Substance amount	
Light intensity	

I found this on page ________.

Express *each measurement in the proper* SI *unit.*

One thousand of the base unit in length: ________________

One millionth of the base unit in mass: ________________

One thousandth of the base unit in time: ________________

Main Idea	Details
I found this on page ______.	**Explain** *models of* accuracy *and* precision.

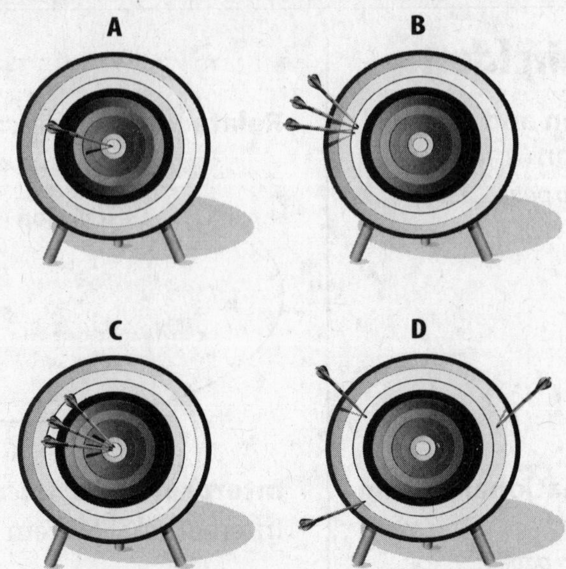

Target	Description	Explanation
		The arrow in the bull's-eye represents one measurement right on the accepted value.
	accurate and precise	
	neither accurate nor precise	
	precise but not accurate	

Measurement and Uncertainty

I found this on page ______.

Relate *two factors that can limit the* accuracy *and* precision *of measurements.*

Factors Affecting Measurement

Main Idea	Details

Significant Digits

I found this on page ________.

Classify *numbers as* significant digits, *or not. Write* S *for significant or* N *for not significant.*

S or N?	Digits
	all nonzero numbers
	zeros used solely for spacing a decimal point
	zeros between nonzero digits
	final zeros after a decimal point

Scientific Tools

I found this on page ________.

Recognize *the uses of scientific tools.*

________ used to record descriptions, explanations, plans, and steps	________ measures the mass of objects	________ measures the temperature of substances
________ used to hold, pour, heat, and measure liquids	________ enables you to see objects too small to view with the eye	________ used to process data

Tools Used by Life Scientists

I found this on page ________.

Describe *tools used by life scientists.*

Tool	Description
Magnifying lens	
Slide	
Dissecting tools	
Pipette	

Connect It Make a generalization about the importance of tools to science.

____

____

____

Lesson 3 Case Study

Skim *Lesson 3 in your book. Read the headings and look at the photos and illustrations. Identify three things you want to learn more about as you read the lesson. Record your ideas in your Science Journal.*

Main Idea	Details
The Biodiesel Revolution *I found this on page ______.*	**Infer** *a primary difference between biodiesel and the main source of energy people have used in industry and transportation for the last few centuries.*
Designing a Controlled Experiment *I found this on page ______.*	**Define** variable, *and express the differences between types of* variables.

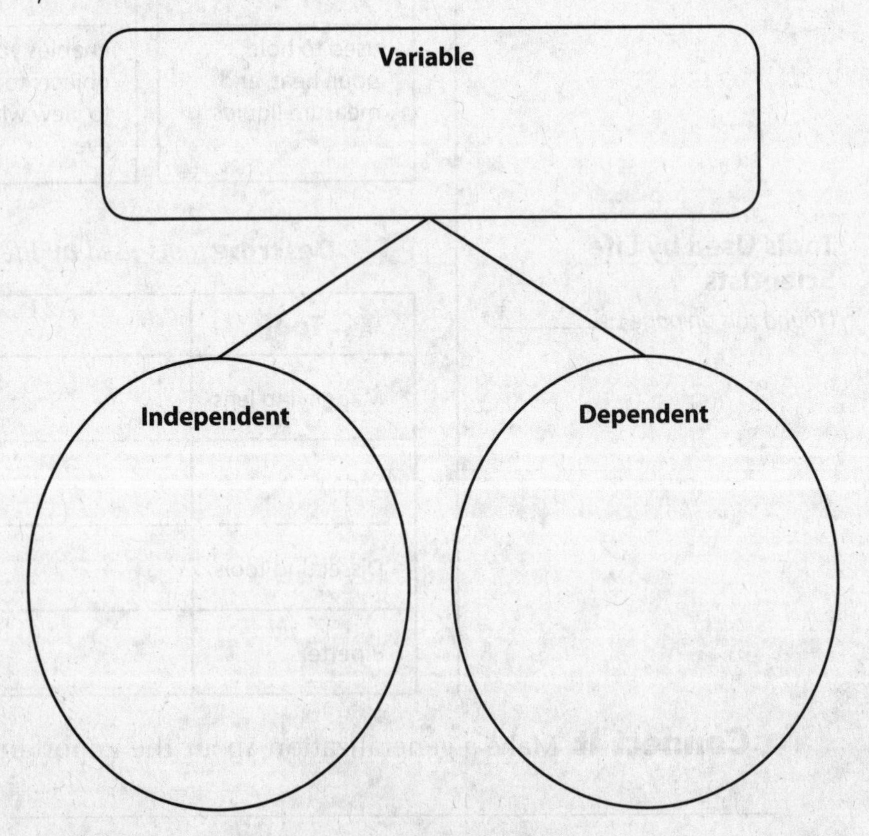

Main Idea	Details
I found this on page ______.	**Identify** *the factors in an experiment that remain the same.*

Main Idea	Details
Biodiesel *I found this on page ______.*	**Assess** *how these two factors affect the preference for biodiesel as a fuel source.* Petroleum: ______ ______ Source of biodiesel: ______ ______ ______
Aquatic Species Program *I found this on page ______.*	**Depict** *the shift in focus of the Aquatic Species Program.* Original focus: → Shifted focus:
Which Microalgae? *I found this on page ______.*	**Record** *a hypothesis formed by scientists evaluating species of microalgae for usefulness in producing biodiesel.* Hypothesis: ______ ______ ______
Oil Production in Algae *I found this on page ______.*	**Evaluate** *the effect that starving microalgae of nitrogen has on oil production.* Less nitrogen → Size of microalgae ______ → Smaller microalgae ______ → Overall oil production ______

Main Idea	Details

Outdoor Testing v. Bioreactors

I found this on page ________.

Contrast *three different growing environments that represent hypotheses about growing microalgae. Identify the major challenge posed by each strategy.*

Growing Environment			
Major Challenge			

Why so many hypotheses?

I found this on page ________.

Restate *what it means for research to be "hypothesis-driven."*

You come up with a question…	

Increasing Oil Yield

I found this on page ________.

Record *a prediction made by scientists seeking to increase the oil yield of microalgae.*

Prediction: ________________________________

____

____

Bringing Light to Microalgae

I found this on page ________.

Identify *2 ways that scientists devised to deliver more light to microalgae to increase productivity of a pond.*

1. ________________________________

____

2. ________________________________

____

Main Idea | Details

Why Grow Microalgae?
I found this on page ________.

Organize *information about the benefits of growing microalgae.*

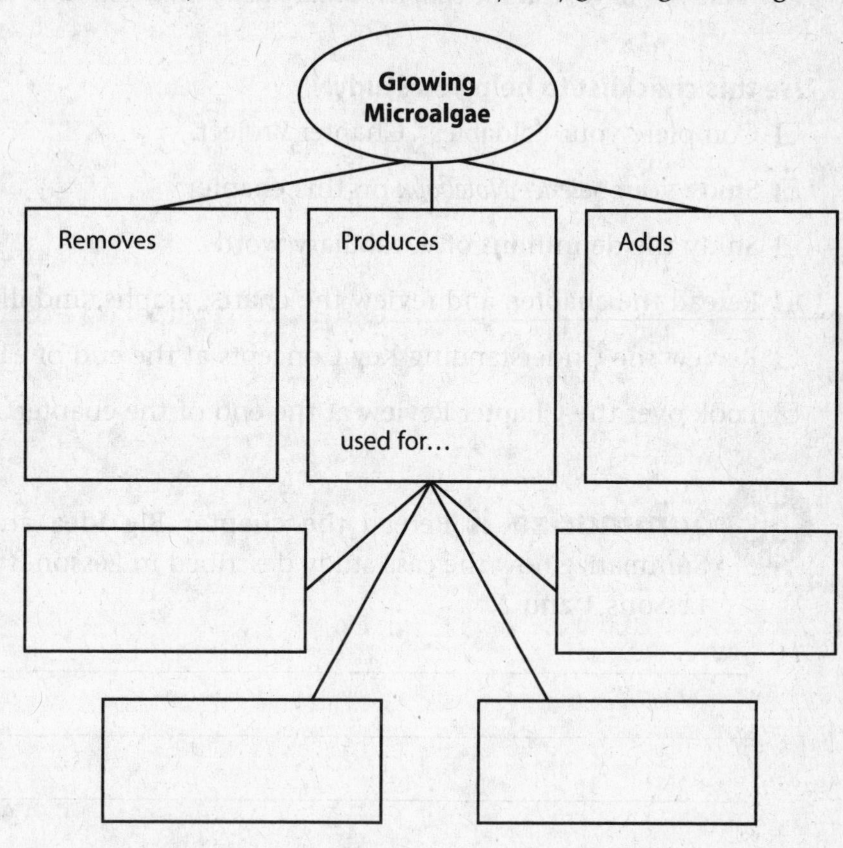

Are microalgae the future?
I found this on page ________.

Conclude *whether biodiesel from microalgae should be the preferred fuel source.*

____

____

____

____

____

____

Synthesize It Identify one hypothesis that was supported and one hypothesis that was not supported throughout the scientific investigation of microalgae as a biodiesel fuel source.

____

____

____

Review

Scientific Explanations

Chapter Wrap-Up

Now that you have read the chapter, think about what you have learned.

Use this checklist to help you study.

- ❑ Complete your Foldables® Chapter Project.
- ❑ Study your *Science Notebook* on this chapter.
- ❑ Study the definitions of vocabulary words.
- ❑ Reread the chapter, and review the charts, graphs, and illustrations.
- ❑ Review the Understanding Key Concepts at the end of each lesson.
- ❑ Look over the Chapter Review at the end of the chapter.

THE BIG IDEA

Summarize It Reread the chapter Big Idea and the lesson Key Concepts. Summarize how the case study described in Lesson 3 relates to the Key Concepts in Lessons 1 and 2.

____

____

____

____

____

____

____

____

____

____

____

____

____

Challenge *Choose another long-term scientific investigation to explore. Research to learn about the problem that scientists are trying to solve. Write a magazine-style article about that real-life application of scientific inquiry. Share your article with your class. (Be sure to avoid personal bias as you write the story!)*

Name ______________________ Date ____________

Classifying and Exploring Life

What are living things, and how can they be classified?

Before You Read

Before you read the chapter, think about what you know about how living things are classified. In the first column, share three things you already know about kinds of living things. In the second column, record three things that you would like to learn more about. When you have completed the chapter, think about what you have learned and complete the **What I Learned** *column.*

K What I Know	W What I Want to Learn	L What I Learned

Chapter Vocabulary

Lesson 1	Lesson 2	Lesson 3
NEW organism cell unicellular multicellular homeostasis	**NEW** binomial nomenclature species genus dichotomous key cladogram	**NEW** light microscope compound microscope electron microscope **ACADEMIC** identify **REVIEW** atom

Lesson 1 Characteristics of Life

Skim *Lesson 1 in your book. Read the headings, and look at the photos and illustrations. Identify three things you want to learn more about as you read the lesson. Write your ideas in your Science Journal.*

Main Idea	Details

Characteristics of Life

I found this on page ______.

Organize *information about living and nonliving things. Complete the word web with the 6 characteristics of life.*

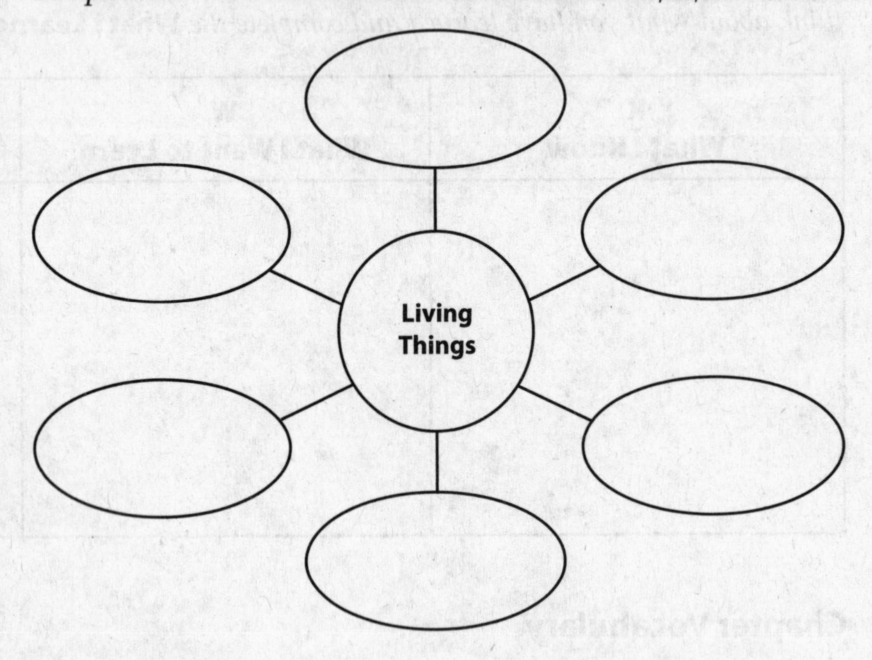

Organization

I found this on page ______.

Describe *the 2 types of organization in* organisms.

1. Unicellular: ______

______

2. Multicellular: ______

______

______

Growth and Development

I found this on page ______.

Compare *growth and development of* multicellular *and* unicellular organisms.

	Multicellular Organism	Unicellular Organism
How the organism grows and develops		

Main Idea	Details

Reproduction

I found this on page ______.

Define *reproduction. Then identify 2 ways in which organisms reproduce.*

Reproduction: ______________________________

____

____

Organisms reproduce by:

1. ______________________________

____

2. ______________________________

____

Responses to Stimuli

I found this on page ______.

Identify *2 types of stimuli, and provide two examples of each.*

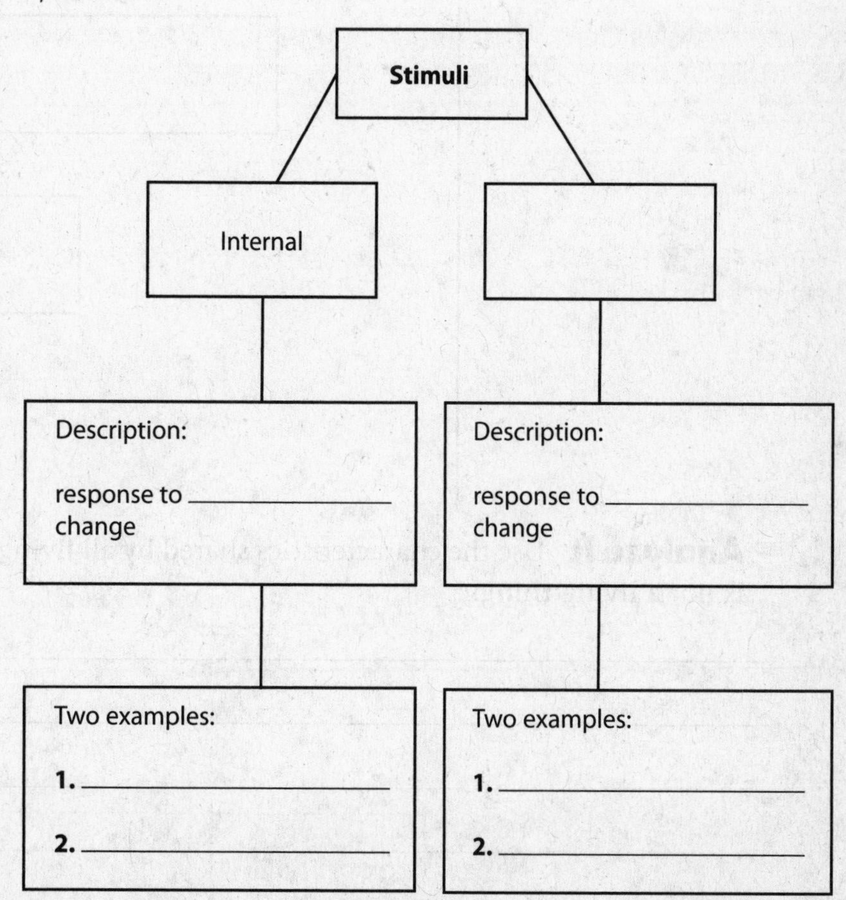

Main Idea	Details

Homeostasis
I found this on page ________.

Analyze *the effect of* homeostasis. *Complete the cause-and-effect chart.*

Cause		Effect
Homeostasis maintained	→	
Homeostasis not maintained	→	

Energy
I found this on page ________.

Sequence *how energy flows from the Sun to a mountain lion.*

Sun → ________ → ________ → ________

Analyze It Use the characteristics shared by all living things to explain why a clock is not a living thing.

____

____

____

____

____

____

Lesson 2 Classifying Organisms

Scan *Lesson 2 in your book. Record three questions you have about classifying living things in your Science Journal. Try to answer your questions as you read.*

Main Idea	Details

Classifying Living Things
I found this on page ______.

Identify *the ways Aristotle organized, or classified, living things.*

Plants	
according to:	according to:
a. ______________ and ______________	**a.** ______________ ______________
b. whether it is ______________, ______________, or ______________	**b.** ______________ and size
	c. ______________

Determining Kingdoms
I found this on page ______.

Indicate *the 5 kingdoms that Whittaker proposed for classifying organisms.*

1. ______________
2. ______________
3. ______________
4. ______________
5. ______________

Determining Domains
I found this on page ______.

Classify *groups of organisms into domains and kingdoms.*

Domain	Kingdom
Bacteria	
	Archaea
Eukarya	
	Plantae

Main Idea	Details
Scientific Names *I found this on page* ______.	**Organize** *information about* binomial nomenclature *by defining each part of a brown bear's scientific name.*

Ursus ***arctus***

Ursus	arctus
Level of classification: ______	**Level of classification:** ______
Description: group of similar species	**Description:** group of organisms ______ ______

I found this on page ______.

Summarize *why scientific names are important.*

______

Classification Tools
I found this on page ______.

Compare *a* dichotomous key *and a* cladogram.

Dichotomous Key	Cladogram

Connect It Compare your first and last names with a scientific name.

______

Lesson 3 Exploring Life

Predict *three things that will be discussed in Lesson 3 after reading the headings. Record your predictions in your Science Journal.*

Main Idea	Details
The Development of Microscopes *I found this on page ______.*	**Describe** *two ways that microscopes have changed people's ideas about living things.*

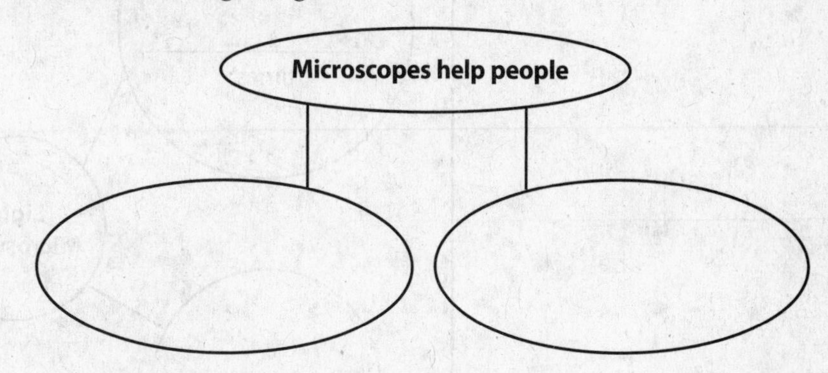

I found this on page ______.

Describe *Anton van Leeuwenhoek's microscope.*

Anton van Leeuwenhoek's microscope was made in the ______________________. The microscope had ______________________ and could magnify an image ______________________. Leeuwenhoek observed ______________________ ______________________ with his microscope.

I found this on page ______.

Explain *what Hooke discovered with his microscope.*

____

____

Types of Microscopes
I found this on page ______.

Identify *2 characteristics of all microscopes.*

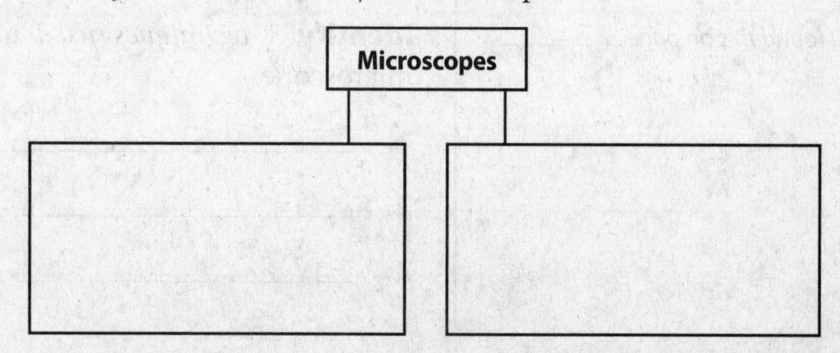

Main Idea	Details
I found this on page ________.	**Organize** *information about* light microscopes *by completing the graphic organizer.*

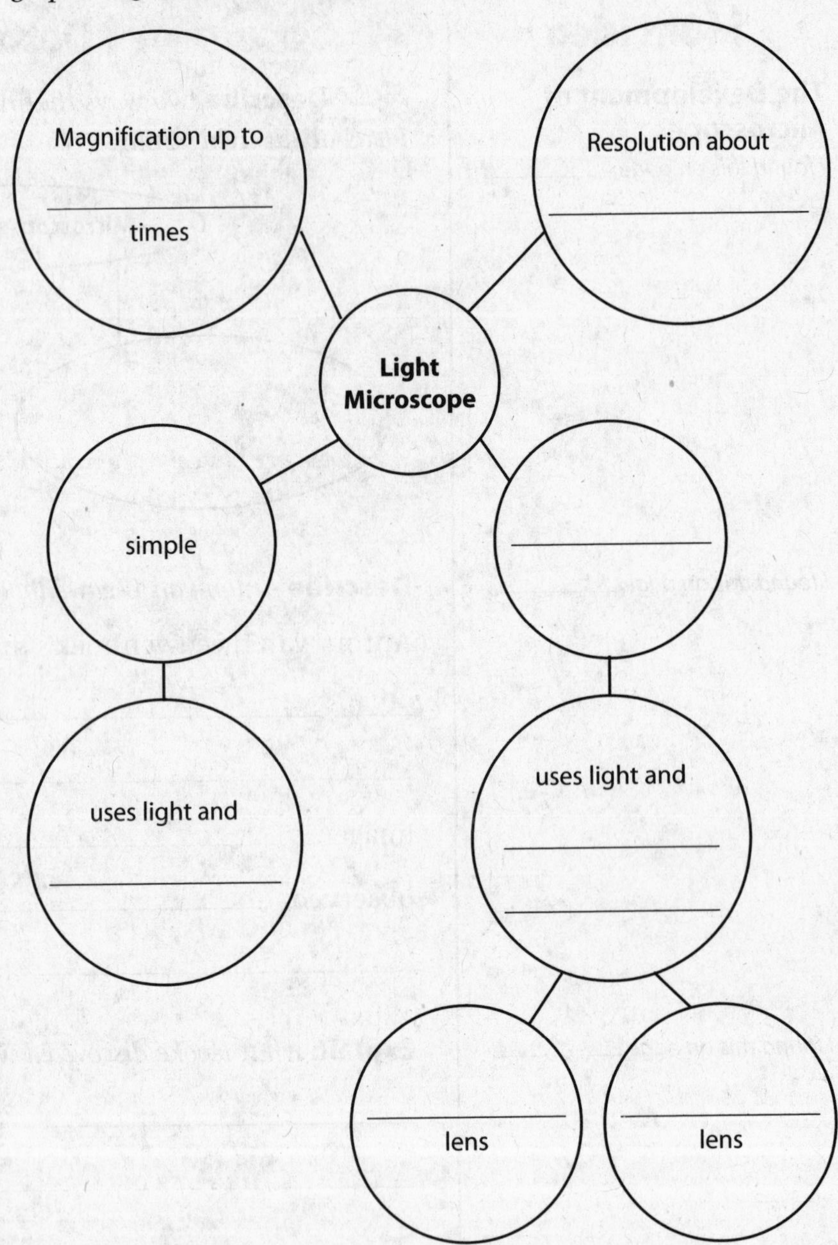

Main Idea	Details
I found this on page ________.	**Identify** *3 techniques used to observe objects with a* light microscope.

1. ______________________________
2. ______________________________
3. ______________________________

Main Idea | Details

Compare and contrast electron microscopes *and* light microscopes.

	Electron Microscope	Light Microscope
Magnification	______ times	______ times
Resolution		
Specimens		
Two Types	1. 2.	1. 2.

I found this on page ______.

I found this on page ______.

I found this on page ______.

Using Microscopes
I found this on page ______

Summarize *the use of microscopes.*

In health care: ______

Other uses: ______

Synthesize It How could you use a light microscope to determine whether spilled crystals were salt or sugar?

Review Classifying and Exploring Life

Chapter Wrap-Up

Now that you have read the chapter, think about what you have learned. Complete the **What I Learned** *column on the first page of the chapter.*

Use this checklist to help you study.

- ❑ Complete your Foldables® Chapter Project.
- ❑ Study your *Science Notebook* on this chapter.
- ❑ Study the definitions of vocabulary words.
- ❑ Reread the chapter, and review the charts, graphs, and illustrations.
- ❑ Review the Understanding Key Concepts at the end of each lesson.
- ❑ Look over the Chapter Review at the end of the chapter.

THE BIG IDEA

Summarize It Review the chapter Big Idea and the lesson Key Concepts. Suppose that you have discovered a new organism. Describe your organism, including whether it is unicellular or multicellular. Tell how Aristotle and Whittaker would have classified the organism. Explain into which domain and kingdom you would place the organism, and why. Then hypothesize how the use of a microscope could help you further describe and classify your organism.

Challenge *Suppose that you found a rock that looks like wood. Describe the process and equipment you would use to determine if the rock was once wood.*

Name ____________________ Date ____________

Cell Structure and Function

How do the structures and processes of a cell enable it to survive?

Before You Read

Before you read the chapter, think about what you know about the topic. Record your thoughts in the first column. Pair with a partner, and discuss his or her thoughts. Record those thoughts in the second column. Then record what you both would like to share with the class in the third column.

Think	Pair	Share

Chapter Vocabulary

Lesson 1	Lesson 2	Lesson 3	Lesson 4
NEW cell theory macromolecule nucleic acid protein lipid carbohydrate **REVIEW** theory	**NEW** cell membrane cell wall cytoplasm cytoskeleton organelle nucleus chloroplast **ACADEMIC** function	**NEW** passive transport diffusion osmosis facilitated diffusion active transport endocytosis exocytosis	**NEW** cellular respiration glycolysis fermentation photosynthesis

Lesson 1 Cells and Life

Skim *Lesson 1 in your book. Read the headings and look at the photos and illustrations. Identify three things you want to learn more about as you read the lesson. Write your ideas in your Science Journal.*

Main Idea	Details
Understanding Cells *I found this on page* ______.	**Explain** *why it took so long for scientists to learn about cells.* ______ ______ ______ ______
I found this on page ______.	**Summarize** *discoveries made by scientists that led to the* cell theory. Robert Hooke ______ ______ ______ Matthias Schleiden ______ ______ ______ Theodor Schwann ______ ______ ______ Rudolf Virchow ______ ______ ______
I found this on page ______.	**List** *the 3 main principles of the* cell theory. **1.** ______ **2.** ______ **3.** ______

Main Idea	Details
Basic Cell Substances *I found this on page* ________.	**Organize** *information about* macromolecules.

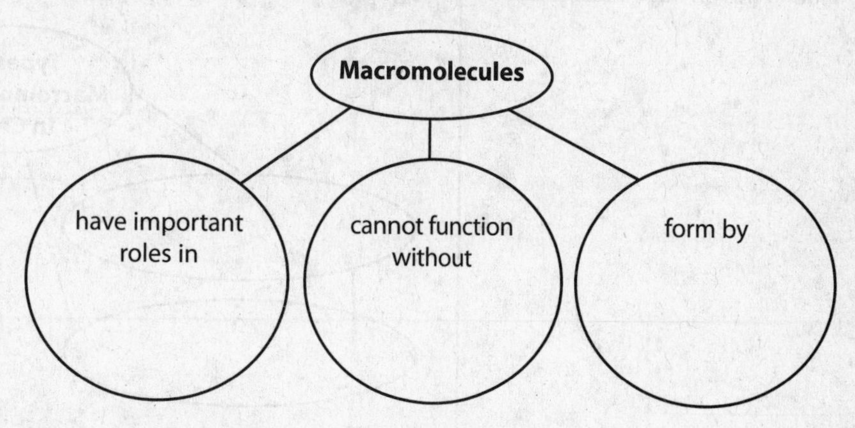

I found this on page ________.

Complete *the statement about basic cell substances.*

The main material inside cells is ____________, which makes up more than ________ percent of the cell's volume.

I found this on page ________.

Draw *a water molecule in the space below. Color the oxygen red and the hydrogen blue, and label the positive and negative ends. In the space below your drawing, describe the structure of the water molecule, and explain:*

1. *how that structure helps dissolve materials;*
2. *why water's ability to dissolve materials is important to the function of a cell.*

____

____

____

____

____

____

Main Idea	Details
I found this on page ______.	**Identify** *the types of* macromolecules *inside cells.*

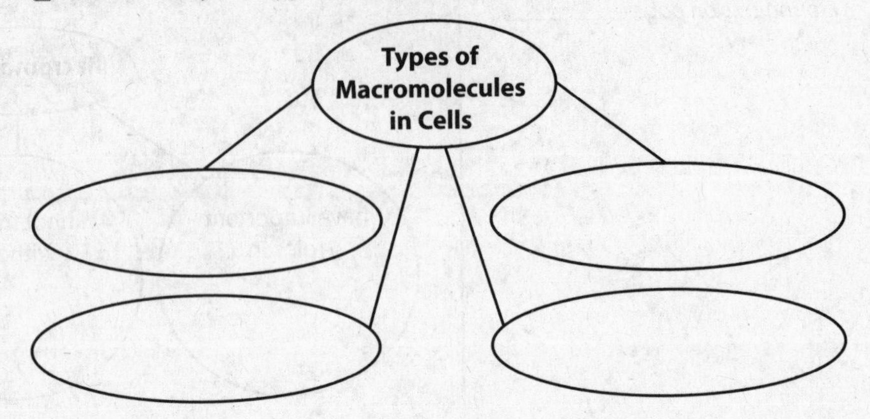

Main Idea	Details
I found this on page ______.	**Distinguish** *2 types of* nucleic acids *and indicate what cells make with each type.* **1.** ______ is used to make ______. **2.** ______ is used to make ______.
I found this on page ______.	**Identify** *4 functions of* proteins. **1.** ______ **3.** ______ **2.** ______ **4.** ______
I found this on page ______.	**Explain** *why* lipids *are able to function as protective barriers in cells.* ______
I found this on page ______.	**Summarize** *information about* carbohydrates.

Carbohydrates	
that provide energy	that provide support
1. ______ 2. ______	1. ______

Connect It Describe how the development of cell theory shows that scientific ideas can change over time. Use specific examples.

______

______

______

______

Lesson 2 **The Cell**

Scan *Lesson 2 in your book. Think of three questions you have about cells. Write those questions in your Science Journal. Then try to answer your questions as you read.*

Main Idea

Cell Shape and Movement

I found this on page ________.

I found this on page ________.

I found this on page ________.

I found this on page ________.

Details

Compare *cell parts by completing the chart. Put a check mark in the* Plant *or* Animal *column to indicate which types of cells contain the cell part listed. You might need to reference the cell diagrams to decide.*

Cell Part		Plant	Animal
Cell membrane	Description: Purpose:		
Cell wall	Description: Purpose:		
Cytoplasm	Description: Purpose:		
Cytoskeleton	Description: Purpose:		

Main Idea	Details

I found this on page ______.

Identify and describe *2 examples of cell appendages.*

1.	Example:	Description: short, hairlike structures	Purpose:
2.	Example: flagella	Description:	Purpose:

Cell Types

I found this on page ______.

Classify *cells as prokaryotic or eukaryotic by writing* "E" *or* "P" *in the right-hand column.*

Characteristic	Cell Type
Cell's genetic material is surrounded by a membrane.	
Cell is usually a unicellular organism.	
It is usually the smaller of the two types of cell.	
Cell contains organelles.	

Cell Organelles

I found this on page ______.

Identify *four facts about* organelles.

1. ______
2. ______
3. ______
4. ______

I found this on page ______.

Describe *some functions of* organelles.

What Organelles Do for Cells

Main Idea | Details

Classify *information about* organelles. *In the right-hand column, indicate whether the* organelle *is in a plant cell, an animal cell, or both.*

Main Idea	Organelle	Function	Plant, Animal, or Both?
I found this on page ________.	Nucleus		
I found this on page ________.	Nucleolus		
I found this on page ________.	Ribosome		
I found this on page ________.	Rough endoplasmic reticulum		
I found this on page ________.	Smooth endoplasmic reticulum		
I found this on page ________.	Mitochondria		
I found this on page ________.	Chloroplast		
I found this on page ________.	Golgi apparatus		
I found this on page ________.	Vesicle		
I found this on page ________.	Central vacuole		
I found this on page ________.	Lysosome		

Synthesize It Some cells contain chloroplasts that use light energy and produce food. Do cells without chloroplasts also depend on sunlight for their food? Explain.

____

____

____

Lesson 3 Moving Cellular Material

Predict *three things that will be discussed in Lesson 3. Read the headings, and look at the photos and illustrations. Write your predictions in your Science Journal.*

Main Idea	Details
Passive Transport *I found this on page* ______.	**List** *2 functions of membranes.* **1.** ______ **2.** ______
I found this on page ______.	**Organize** *information about* passive transport.

Passive Transport

Definition:	Depends on:	Example:

Main Idea	Details
Diffusion *I found this on page* ______.	**Assess** *information about* diffusion. *Read the statements below. If the statement is true, write* true *on the line. If it is false, rewrite the underlined portion of the statement so that it is true.*

Diffusion is the movement of substances from an area of <u>lower concentration</u> to an area of <u>higher concentration.</u>

______

Diffusion continues until the concentration of a substance is <u>higher inside a cell than outside a cell.</u>

______

Main Idea	Details
Osmosis—The Diffusion of Water *I found this on page* ______.	**Complete** *the sentence about* osmosis. Osmosis is a type of ______________ that involves movement of ______________ only through the cell membrane.
I found this on page ______.	**Explain** *the process of* facilitated diffusion.

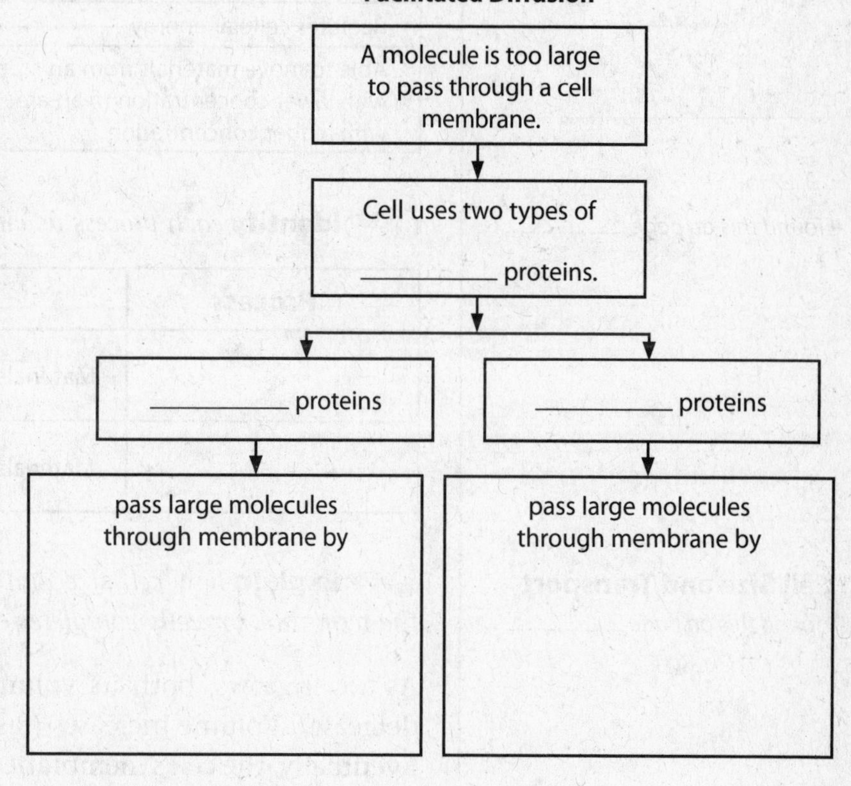

Active Transport
I found this on page ______.

Organize *information about* active transport.

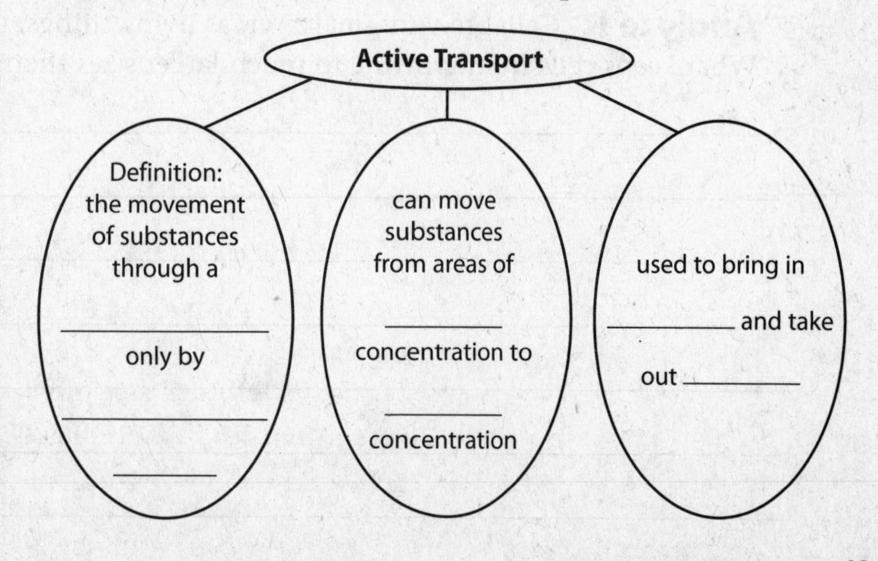

Main Idea | Details

I found this on page ______.

Compare and contrast facilitated diffusion *and* active transport *by writing* yes *or* no *in each empty box of the chart.*

Description	Facilitated Diffusion	Active Transport
Uses carrier proteins		
Transports materials across cell membrane		
Requires cellular energy		
Able to move materials from an area with lower concentration to an area with higher concentration		

I found this on page ______.

Identify *each process as either* endocytosis *or* exocytosis.

Process	Description
	Materials entering cell
	Materials being expelled from cell

Cell Size and Transport
I found this on page ______.

Explain *how cell size and transport are related. Underline the term that correctly completes each sentence.*

As a cell grows, both its volume and surface area (increase/decrease). Volume increases (faster/slower) than surface area. Eventually, the cell's membrane would be (too large/too small) to move enough materials into and out of the cell.

Analyze It Cells are very small. Yet, as living things, they have the ability to grow. What keeps cells from growing to much larger sizes than they do?

Lesson 4 Cells and Energy

Scan *Lesson 4 in your book. Think of three questions you have about cells and energy. Write those questions in your Science Journal. Then try to answer your questions as you read.*

Main Idea	Details
Cellular Respiration *I found this on page ______.*	**Organize** *information about* cellular respiration.

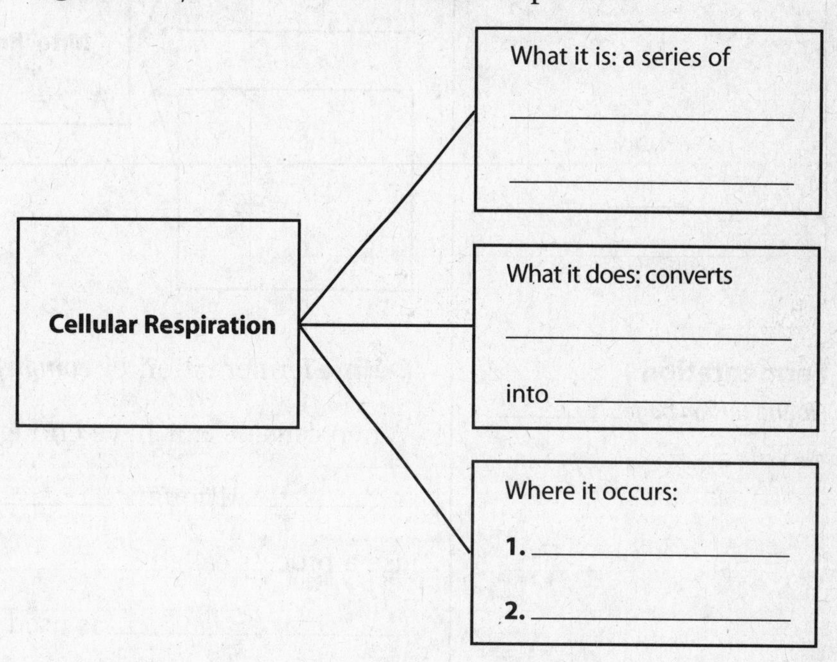

I found this on page ______.

Summarize *the first step in the process of* cellular respiration. *Then label the steps in the diagram on the left.*

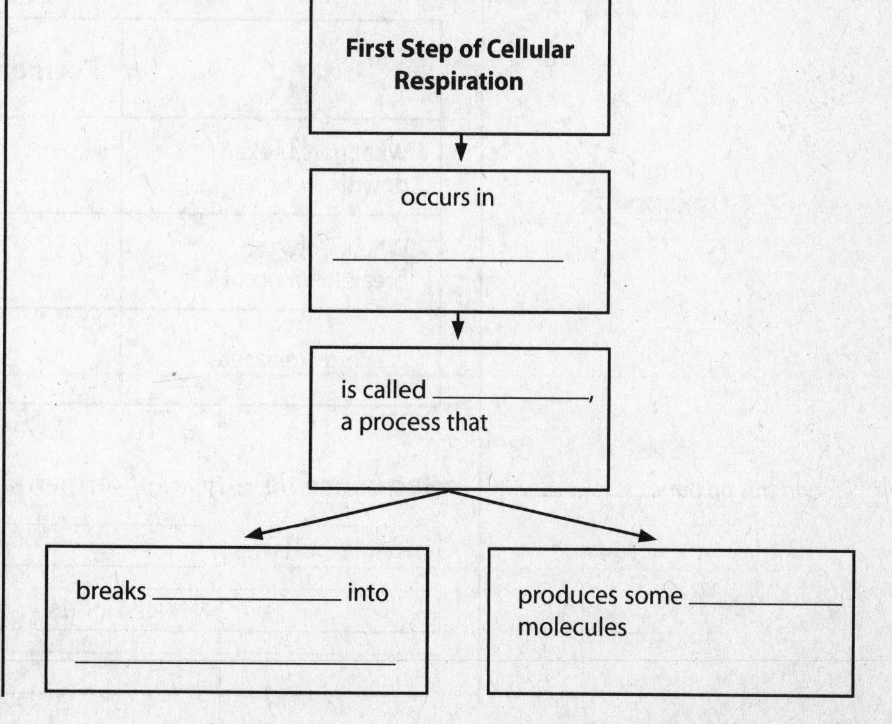

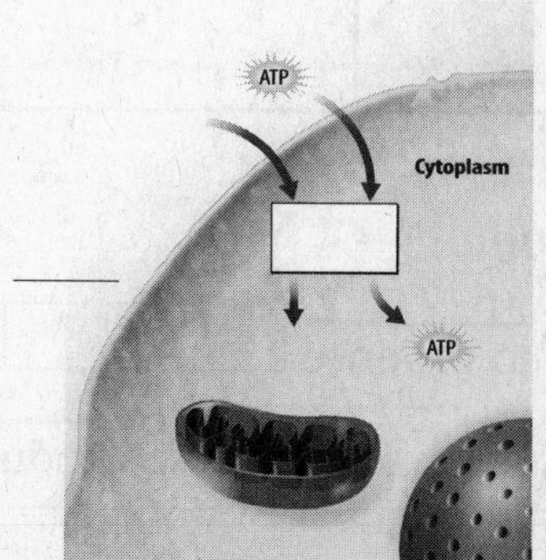

Main Idea | Details

I found this on page ______.

Describe *the second step of* cellular respiration.

Fermentation

I found this on page ______.

Define fermentation *by completing the sentences.*

When cells do not have enough ______ to make ______ through ______, they use a process called ______. Because no ______ is used, less ______ is produced than in ______.

I found this on page ______.

Compare fermentation *to* cellular respiration.

	Fermentation	Cellular Respiration
What gets broken down?		
Where does the breakdown occur?		
Is energy released?		

I found this on page ______.

Sequence *the 2 types of* fermentation.

muscle cells use ______ → to produce ______ + ______

yeast cells use ______ → to produce ______ + ______ + ______

Main Idea	Details
Photosynthesis *I found this on page* ______.	**Diagram** *the reactions that occur in chloroplasts during* photosynthesis *in the space below. Show what goes into and comes out of this process. Use these terms:* • sugar • oxygen • light energy • water • carbon dioxide
I found this on page ______.	**Create** *a cycle diagram that shows the relationship between* photosynthesis *and* cellular respiration. *Use the terms* chloroplast, glucose, oxygen, water, carbon dioxide, light energy, *and* mitochondrion *in your model.*

Analyze It Why is photosynthesis important to living things other than plants?

____

____

____

____

Review Cell Structure and Function

Chapter Wrap-Up

Now that you have read the chapter, think about what you have learned.

Use this checklist to help you study.

- ❑ Complete your Foldables® Chapter Project.
- ❑ Study your *Science Notebook* on this chapter.
- ❑ Study the definitions of vocabulary words.
- ❑ Reread the chapter, and review the charts, graphs, and illustrations.
- ❑ Review the Understanding Key Concepts at the end of each lesson.
- ❑ Look over the Chapter Review at the end of the chapter.

THE BIG IDEA

Summarize It *Reread the chapter Big Idea and the lesson Key Concepts. When scientists first began to study cells, they found that plant and animal cells were similar to each other and yet different from each other. What are the similarities that you have noticed? What are the differences?*

Challenge *Compare the cell to a factory. For example, a factory has a manager, and a cell has a nucleus. Use similar analogies to describe the functions of different parts of the cell.*

Name ______________________ Date ______________

From a Cell to an Organism

How can one cell become a multicellular organism?

Before You Read

Before you read the chapter, think about what you know about cells. Record your thoughts in the first column. Pair with a partner, and discuss his or her thoughts. Record those thoughts in the second column. Then record what you both would like to share with the class in the third column.

Think	Pair	Share

Chapter Vocabulary

Lesson 1	Lesson 2
NEW cell cycle interphase sister chromatids centromere mitosis cytokinesis daughter cells **REVIEW** eukaryotic	**NEW** cell differentiation stem cell tissue organ organ system **ACADEMIC** complex

Lesson 1 The Cell Cycle and Cell Division

Scan *Lesson 1. Read the headings and the bold words. Look at the pictures. Identify three facts you discovered about the cell cycle and cell division. Record them in your Science Journal.*

Main Idea	Details
The Cell Cycle *I found this on page ______.*	**Explain** *the* cell cycle. Cell cycle: ______________________ ______________________
I found this on page ______.	**Organize** *information about the 2 main phases of the* cell cycle.

Main Phases of the Cell Cycle

Interphase	Mitotic Phase
Duration: longer shorter	Duration: longer shorter
Description: ______	Description: ______

I found this on page ______.

Complete *the sentence to explain why the length of a* cell cycle *varies.*

The length of a cell cycle depends on

____

______________________.

Main Idea	Details

Interphase

I found this on page ______.

Represent *the relative length of each stage of* interphase *by labeling the time line.*

Describe *each stage of* interphase.

Stages of Interphase	
Stage	**Description**
______	______ and ______ functions
S	______ and ______ replication; The two new strands of DNA are called ______ and are held together by the ______.
______	______ and preparation for ______

I found this on page ______.

Assess *information about organelle replication. Read the statement below. If the statement is true, write* true *on the line. If it is false, rewrite the underlined portion of the statement so that it is true.*

Organelle replication occurs <u>only during the S stage</u> of interphase. ______

I found this on page ______.

Organize *information to describe the stages in the mitotic phase of the* cell cycle.

Mitotic Phase of the Cell Cycle

- Mitosis
  - Description: ______
- ______
  - Description: ______

Main Idea | **Details**

The Mitotic Phase
I found this on page ______.

Identify *each phase of* mitosis.

______
DNA condenses; spindle fibers begin to form.

I found this on page ______.

______
Chromosomes line up in a single file at the middle of the cell.

I found this on page ______.

______
Sister chromatids separate and pull to opposite sides.

I found this on page ______.

______
Nuclear membrane reforms; chromosomes unwind.

I found this on page ______.

Describe cytokinesis *in plants and animals.*

In animals: ______

______

In plants: ______

______

Results of Cell Division
I found this on page ______.

Summarize *4 results of the* cell cycle.

1. Reproduction	
2.	allows multicellular organisms to grow from one cell to many
3.	replaces worn-out or damaged cells with new cells
4. Repair	

Connect It Apply what you have learned to explain what probably happened when the bean plant grew overnight in the story of *Jack and the Beanstalk.*

______

______

______

Levels of Organization

Predict *three facts that will be discussed in Lesson 2 after reading the headings. Write your predictions in your Science Journal.*

Main Idea	Details
Life's Organization *I found this on page ______.*	**Summarize** *life's organization.* All organisms ______________________.
Unicellular Organisms *I found this on page ______.*	**Organize** *information about unicellular organisms by completing the graphic organizer.*

Unicellular Organisms

____

Description:
- cell without a membrane-bound ______________
- smaller than ______________; some live in ______________ ______________ ______________

____

Description:
- cell has a ______________ surrounded by a ______________
- many specialized ______________ ______________

____

- obtain ______________
- respond ______________
- grow
- ______________ ______________ ______________

Main Idea	Details
Multicellular Organisms *I found this on page ________.*	**Compare and contrast** *unicellular and multicellular organisms.*

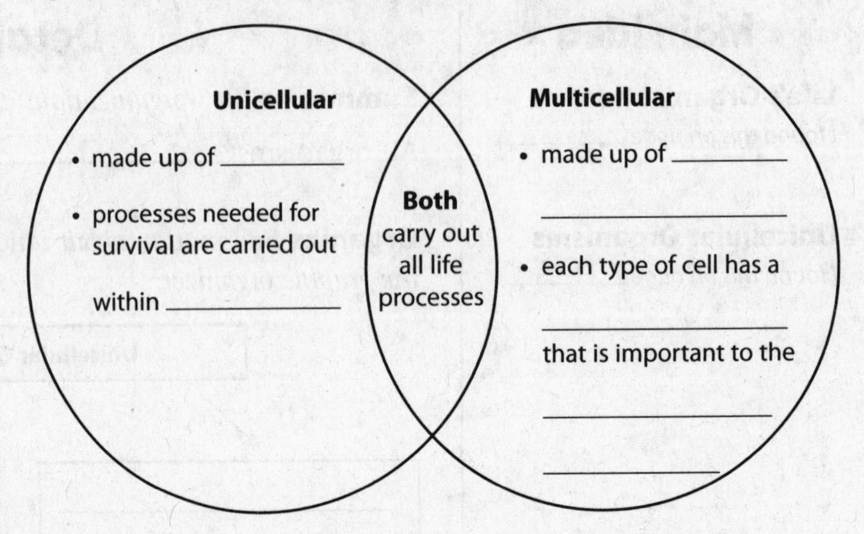

I found this on page ________.

Organize *information about* cell differentiation.

Cell Differentiation

Definition: ________________________

In plants (meristems)	In animals (stem cells)
located:	in embryos:
what it does:	in adults:

Main Idea | Details

I found this on page ________.

Identify *the 4 main types of animal* tissue.

1. ________ **3.** ________

2. ________ **4.** ________

I found this on page ________.

Identify *3 main types of plant* tissue, *and tell the function of each.*

Plant Tissue	
Type	**Function**
1. Dermal	
2.	
3. Ground	

I found this on page ________.

Sequence *the organization of cells,* tissues, organs, *and* organ systems *in a multicellular organism.*

Cells are organized in ________.

Different ________ working together to perform a particular job are called ________.

Groups of ________ that work together to complete a series of tasks are called ________.

Many ________ working together make up an ________.

Connect It The cells of all your organs have the same DNA in their nuclei, yet all perform different jobs in your body. Explain how this can be so. Use the term *cellular differentiation* in your explanation.

____

____

____

Review

From a Cell to an Organism

Chapter Wrap-Up

Now that you have read the chapter, think about what you have learned.

Use this checklist to help you study.

- ❑ Complete your Foldables® Chapter Project.
- ❑ Study your *Science Notebook* on this chapter.
- ❑ Study the definitions of vocabulary words.
- ❑ Reread the chapter, and review the charts, graphs, and illustrations.
- ❑ Review the Understanding Key Concepts at the end of each lesson.
- ❑ Look over the Chapter Review at the end of the chapter.

Summarize It Reread the chapter Big Idea and the lesson Key Concepts. Draw the 4 phases of mitosis, and label your drawing. Tell how mitosis is important for both unicellular and multicellular organisms.

Challenge *Unicellular organisms are sometimes called "simple" organisms. Imagine that you are involved in a debate and must argue against this description. What would you say?*

Name ______________________________ Date ______________

Reproduction of Organisms

THE BIG IDEA Why do living things reproduce?

Before You Read

Before you read the chapter, think about what you know about why living things reproduce. In the first column, share three things you already know about this topic. In the second column, identify three things that you would like to learn more about. When you have completed the chapter, think about what you have learned and complete the **What I Learned** *column.*

K What I Know	W What I Want to Learn	L What I Learned

Chapter Vocabulary

Lesson 1	Lesson 2
NEW sexual reproduction egg sperm fertilization zygote diploid homologous chromosomes haploid meiosis **REVIEW** DNA	**NEW** asexual reproduction fission budding regeneration vegetative reproduction cloning **ACADEMIC** potential

Lesson 1 Sexual Reproduction and Meiosis

Scan *Lesson 1. Then write three questions you have about sexual reproduction in your Science Journal. Try to answer your questions as you read.*

Main Idea / Details

What is sexual reproduction?
I found this on page ______.

Model *the process of* sexual reproduction. *Complete the diagram using these labels:*

- egg
- sperm
- fertilization
- zygote

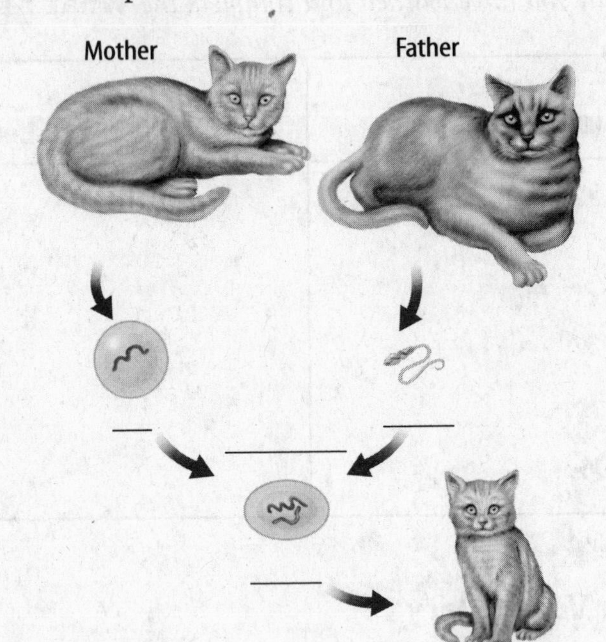

Diploid Cells
I found this on page ______.

Detail *the relationship between* diploid *cells and* homologous chromosomes.

____

____

____

____

Haploid Cells
I found this on page ______.

Define haploid *cells, and explain how they are produced.*

____

____

____

____

Main Idea	Details

The Phases of Meiosis

I found this on page ______.

Model *the stages of* meiosis *I. Draw and describe each stage.*

Stage of Meiosis I	Drawing	Description
Prophase I		
Metaphase I		
Anaphase I		
Telophase I		

I found this on page ______.

Model *the stages of* meiosis *II. Describe each stage.*

Stage of Meiosis II	Description
Prophase II	
Metaphase II	
Anaphase II	
Telophase II	

Main Idea | Details

Why is meiosis important?

I found this on page ______.

Summarize *the importance of* meiosis.

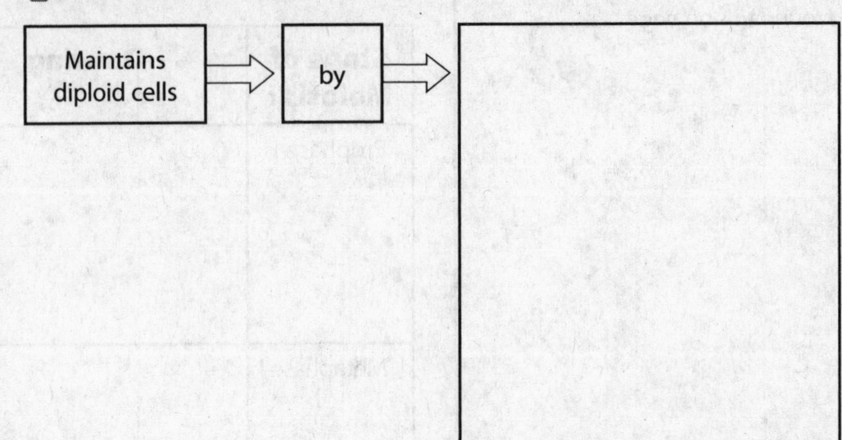

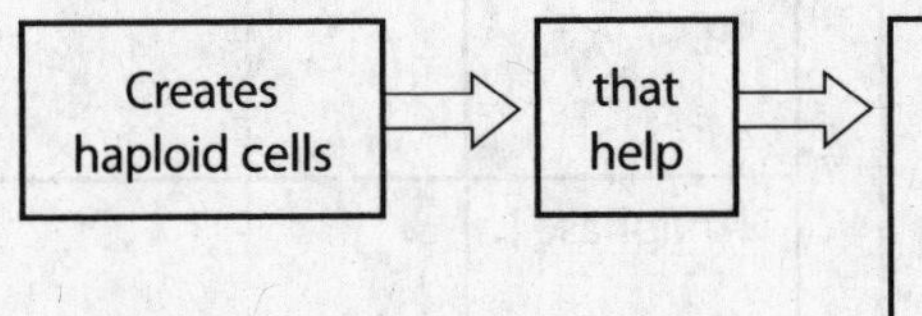

How do mitosis and meiosis differ?

I found this on page ______.

Compare and contrast meiosis *and* mitosis *and cell division.*

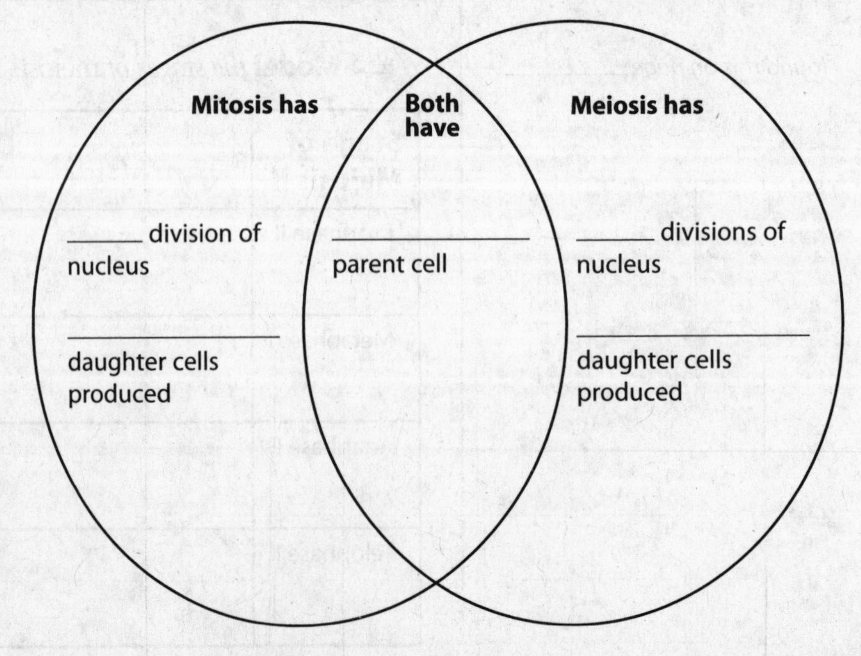

Main Idea	Details

Advantages of Sexual Reproduction
I found this on page ______.

Explain *why genetic variation and selective breeding are advantages of* sexual reproduction.

Advantage	Explanation
Genetic variation	
Selective breeding	

Disadvantages of Sexual Reproduction
I found this on page ______.

Identify *two main disadvantages of* sexual reproduction.

1. ______________________________

2. ______________________________

Connect It Explain how the process of meiosis relates to the way in which a child resembles but is not an exact copy of his or her parents.

____

____

____

____

____

____

____

____

____

____

Lesson 2 Asexual Reproduction

Predict *three facts that will be discussed in Lesson 2 after reading the headings. Write your facts in your Science Journal.*

Main Idea	Details

What is asexual reproduction?

I found this on page ________.

Identify *key points about* asexual reproduction. *Cross out the terms that do not apply to the process.*

single parent organism	fertilization	genetically identical	meiosis
diploid parent cells	offspring produced	2 parent organisms	haploid daughter cells

I found this on page ________.

Summarize asexual reproduction *in your own words.*

____

____

____

____

Types of Asexual Reproduction

I found this on page ________.

I found this on page ________.

I found this on page ________.

I found this on page ________.

I found this on page ________.

Identify *the 6 types of* asexual reproduction.

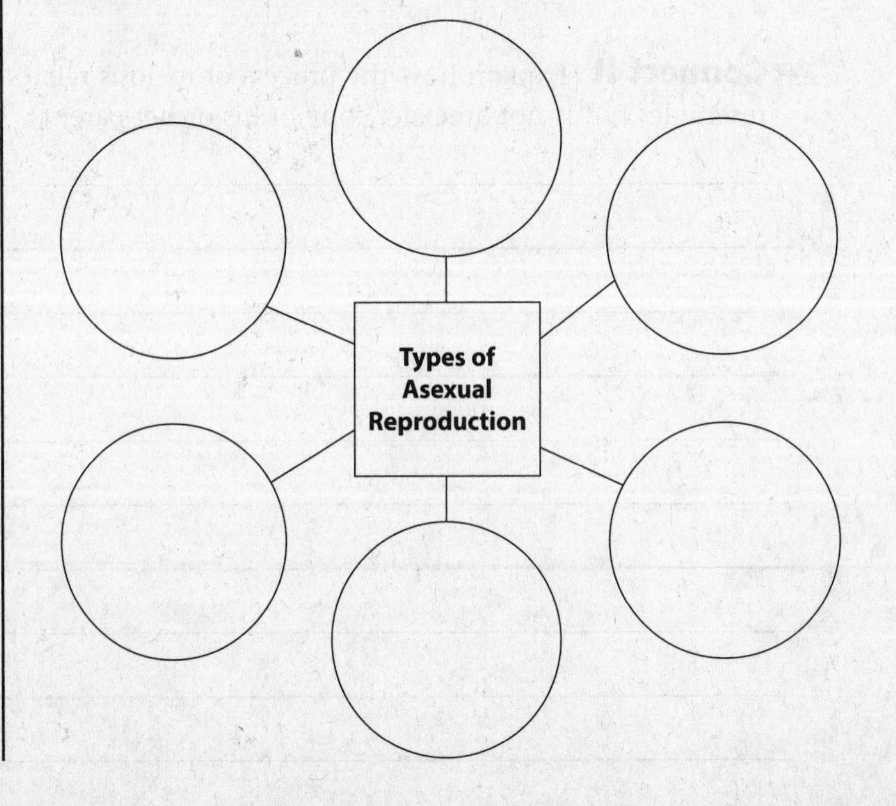

Main Idea	Details
I found this on page ______.	**Sequence** *the steps of cell division through* fission.

1. Fission starts with a prokaryote, which does not have a membrane-bound nucleus.
2. The prokaryote's ______________________ is copied.
3. The cell grows longer, pulling the two ______________ apart.
4. The cell membrane ______________________ ______________________.
5. The cell splits. Two ______________________ are formed.

I found this on page ______.

Write *a complete sentence that defines mitotic cell division and identifies what type of organism undergoes the process.*

____

____

____

____

I found this on page ______.

Draw *a representation of* budding. *Write a definition of the term on the lines below your drawing.*

Definition: ______________________________________

____

____

____

Main Idea	Details
I found this on page ______.	**Explain** *how animal* regeneration *can produce two results.*

Animal regeneration produces

- new ______.
  - A complete offspring ______ ______ ______.
- new ______ organisms.
  - An organism can grow a ______ when ______ ______.

Main Idea	Details
I found this on page ______.	**Identify** *the structures of plants usually involved in* vegetative reproduction.

______ ______ ______

Main Idea	Details
I found this on page ______.	**Explain** *how the definition of* cloning *has changed over time.*

Cloning

In the past	Today

Main Idea	Details
I found this on page ______.	**Identify** *three advantages of using tissue culture to clone plants.*

1. ______

2. ______

3. ______

Main Idea	Details

I found this on page ________.

Sequence *the steps scientists used to produce the cloned sheep, Dolly.*

1. A cell is removed from the first animal. DNA is removed from an unfertilized egg cell from a second animal.
2. The cells from the two animals are ____________. The new cell contains ____________________.
3. The cell develops into an embryo in the lab.
4. ____________________ into the animal that donated the unfertilized egg.
5. A new individual is born. This individual is an ____________ ____________________.

I found this on page ________.

Classify *features of* asexual reproduction *as advantages or disadvantages. Write "A" for advantage and "D" for disadvantage in the center column of the table below. Explain your reasoning in the right-hand column.*

Does not require a mate		
Can occur rapidly		
Produces little genetic variation		

Synthesize It Use your understanding of asexual reproduction to explain why it is important that organisms reproduce in a variety of ways.

____

____

____

____

Review

Reproduction of Organisms

Chapter Wrap-Up

Now that you have read the chapter, think about what you have learned. Complete the **What I Learned** *column on the first page of the chapter.*

Use this checklist to help you study.

- ❑ Complete your Foldables® Chapter Project.
- ❑ Study your *Science Notebook* on this chapter.
- ❑ Study the definitions of vocabulary words.
- ❑ Reread the chapter, and review the charts, graphs, and illustrations.
- ❑ Review the Understanding Key Concepts at the end of each lesson.
- ❑ Look over the Chapter Review at the end of the chapter.

THE BIG IDEA

Summarize It Reread the chapter Big Idea and the lesson Key Concepts. Imagine how the human population would be different if humans reproduced asexually. Explain how this could be both an advantage and a disadvantage to humans and to other organisms.

Challenge *Design two models that demonstrate how genetic material is passed from parents to offspring in meiosis and in mitotic cell division. Present your models to the class, and explain the processes that they represent.*

Name ______________________________ Date ______________

Genetics

How are traits passed from parents to offspring?

Before You Read

Before you read the chapter, think about what you know about genetics. In the first column, record three things you already know about the passage of traits from parents to offspring. In the second column, write three things you would like to learn about this topic. When you have completed the chapter, think about what you have learned and complete the **What I Learned** *column.*

K What I Know	W What I Want to Learn	L What I Learned

Chapter Vocabulary

Lesson 1	Lesson 2	Lesson 3
NEW heredity genetics dominant trait recessive trait **REVIEW** sperm egg	**NEW** gene allele phenotype genotype homozygous heterozygous Punnett square incomplete dominance codominance polygenic inheritance **ACADEMIC** conclude	**NEW** DNA nucleotide replication RNA transcription translation mutation

Lesson 1 Mendel and His Peas

Scan *Lesson 1. Read the lesson titles and bold words. Look at the pictures. Identify three facts that you discovered about Mendel and his peas. Record your facts in your Science Journal.*

Main Idea	Details
Early Ideas About Heredity *I found this on page* ______.	**Define** heredity. Heredity: ______________________
I found this on page ______.	**Describe** genetics, *and explain why Gregor Mendel is known as the father of* genetics. ______________________
Mendel's Experimental Methods *I found this on page* ______.	**Identify** *three reasons why Mendel chose pea plants for his genetic studies.* **1.** ______ **2.** ______ **3.** ______
I found this on page ______.	**Sequence** *the 2 ways pea plants pollinate.* ______ + ______ = ______ from the same plant; ______ + ______ = ______ from different plants
I found this on page ______.	**Detail** *the 2 ways in which cross-pollination can occur.*

Cross-Pollination

Occurs naturally by	Occurs artificially by
1.	**1.**
2.	
3.	

Main Idea — Details

I found this on page ______.

Define *true-breeding plants.*

True-breeding plants → always produce → ______

I found this on page ______.

Sequence *the steps in Mendel's cross-pollination process.*

1. Mendel removed the ______ from the ______.
2. He ______ from the ______ of the white flower to the ______ of the purple flower.
3. He planted ______ that ______ of the purple flower.
4. ______

I found this on page ______.

Explain *why Mendel performed these experiments.*

______

______

______

______

Mendel's Results

I found this on page ______.

Model *the results of Mendel's first generation crosses.*

(P) × (P) = ______

(W) × (W) = ______

(P) × (W) = ______

I found this on page ______.

Write *two questions raised by Mendel's first-generation cross results.*

1. ______

______

2. ______

______

Main Idea	Details
I found this on page ______.	**Model** *the results of Mendel's second-generation (hybrid) cross. Describe Mendel's results.*
	$(P_{hybrid}) \times (P_{hybrid}) =$
I found this on page ______.	Result:
I found this on page ______.	**Explain** *the pattern produced when Mendel crossed two hybrids for a given trait.*
	3:1 ______
Mendel's Conclusions *I found this on page* ______.	**Summarize** *Mendel's conclusions.*
I found this on page ______.	**Explain** *how* dominant *and* recessive *factors interact.*

Dominant Factor	Recessive Factor

Synthesize It How would you determine if wrinkled leaves or smooth leaves is the dominant factor in a true-breeding plant?

Lesson 2 Understanding Inheritance

Predict *three facts that will be discussed in Lesson 2 after reading the headings. Record your facts in your Science Journal.*

Main Idea	Details
What controls traits? *I found this on page* ______.	**Record** *three facts about chromosomes.* **1.** ______ **2.** ______ **3.** ______ ______
I found this on page ______.	**Define** gene *and* alleles. Gene: ______ ______ Alleles: ______
I found this on page ______.	**Distinguish** genotype *and* phenotype.
I found this on page ______.	**Explain** *how symbols are used to represent* alleles *in a* genotype. ______ ______
I found this on page ______.	**Identify** *whether each* genotype *exhibits the recessive or dominant* phenotype. *Circle the* genotypes *that are* homozygous. *Rr:* ______ *rr:* ______ *RR:* ______

Phenotype	Genotype

Main Idea	Details
Modeling Inheritance	**Identify** *the 2 models used to identify and predict traits among genetically related individuals.*
I found this on page ________.	**1.** ________________ **2.** ________________
I found this on page ________.	**Define** Punnett square.
I found this on page ________.	**Complete** *and analyze the* Punnett square. *R represents round seeds; r represents wrinkled seeds.*

	R	*r*
R	*RR*	
r		

Ratio of phenotypes: ________________

Ratio of genotypes: ________________

Percent of offspring with round seeds: ________________

Percent of offspring with wrinkled seeds: ________________

I found this on page ________.

Draw *a pedigree chart that reflects this information: Two parents have five children. Both of the parents have curly hair. Two boys and one girl have curly hair; the other two have straight hair. Before you draw your chart, choose a color for straight and curly hair, and indicate it in the table. After you draw your chart, determine which trait is dominant and label the proper columns in the table.*

________ phenotype	________ phenotype
⬭ Female with curly hair	⬭ Female with straight hair
▭ Male with curly hair	▭ Male with straight hair

Main Idea	Details

Complex Patterns of Inheritance

I found this on page ______.

Define *2 types of* dominance.

Incomplete dominance: ______________________

____

Codominance: ______________________

____

I found this on page ______.

Explain *the idea of multiple* alleles.

____

____

____

I found this on page ______.

Compare *Mendel's interpretation of inheritance with* polygenic inheritance.

Mendel's Conclusion	Polygenic Inheritance
Each trait is determined by ______________.	______________ determine a trait, so many ______________ affect the phenotype.

Genes and the Environment

I found this on page ______.

Analyze *2 factors that can affect an organism's* phenotype.

Genes plus ______ and ______ affect phenotype.

Connect It Eye color is a trait determined by polygenic inheritance. Explain whether using a pedigree chart would work to trace this trait in a family.

____

____

____

____

____

Lesson 3 DNA and Genetics

Skim *Lesson 3 in your book. Read the headings and look at the photos and illustrations. Identify three things you want to learn more about as you read the lesson. Record your ideas in your Science Journal.*

Main Idea	Details
The Structure of DNA *I found this on page* ______.	**Define** DNA, *and explain the relationship of* DNA *and genes.* ______________________ ______________________
I found this on page ______.	**Describe** *the shape of a* DNA *molecule.* ______________________
I found this on page ______.	**Identify** *the 3 components of a* nucleotide, *and tell where each component is found in a* DNA *molecule.*

A nucleotide is a molecule made of

In DNA:	In DNA:	In DNA:

Main Idea	Details
I found this on page ______.	**Identify** *the 4 nitrogen bases found in* DNA, *and finish the statement about the nitrogen bases.* **1.** __________ **3.** __________ **2.** __________ **4.** __________ __________ always bonds to T; __________ always bonds to G.
I found this on page ______.	**Sequence** *the* DNA replication *process.*

1. DNA strand separates and ______________.
2. Nucleotides form new ______________.
3. Two ______________ are produced.

Main Idea	Details

Making Proteins

I found this on page ______.

Explain *the role* DNA *plays in making proteins.*

DNA

↓

carries a complete set of ______ that provide ______ for all the ______ that a cell needs.

↓ ↓

Most genes contain:

Some genes contain:

I found this on page ______.

Explain *the term junk* DNA *and its function.*

______

______

I found this on page ______.

Define RNA, *and describe its 2 functions.*

RNA, or ribonucleic acid, is a type of ______. It

1. ______, and

2. ______.

I found this on page ______.

Compare DNA *and* RNA.

	RNA	DNA
Made of		
Number of strands		
Nitrogen base		
Sugar		

Main Idea	Details
I found this on page ______.	**Describe** transcription. Transcription: ______________________
I found this on page ______.	**Sequence** *the 2 steps involved in* transcription. **1.** ______________________ **2.** ______________________
I found this on page ______.	**Identify** *3 types of* RNA *and their abbreviations.* **1.** ______________________ **2.** ______________________ **3.** ______________________
I found this on page ______.	**Define** translation, *and tell where this process occurs.* ______________________ ______________________
I found this on page ______.	**Sequence** *the process of* translation. **1.** ______ carries amino acids to the ______. ↓ **2.** ______ helps form chemical bonds that ______. ↓ **3.** The first ______ separates from its amino acid and from the ______. A third ______ brings in another ______.
I found this on page ______.	**Explain** *the part that codons play in making proteins.* ______________________ ______________________
Mutations *I found this on page* ______.	**Define** mutation. *Identify two factors that can trigger them.* Mutations: ______________________ ______________________ Triggered by: **1.** ______ **2.** ______

Main Idea	Details
I found this on page ______.	**Analyze** *3 types of* mutations. **1.** Deletion: ______ **2.** Insertion: ______ **3.** Substitution: ______
I found this on page ______.	**Identify** *the effects of a* mutation.

Cause		Effect
Proteins express traits. Mutations change ______, which changes ______.	→	Changed traits can ______ or ______ an organism.

Identify *four genetic disorders caused by* mutations.

1. ______

2. ______

3. ______

4. ______

Connect It Are genetic disorders always inherited? Explain your answer.

______

Review Genetics

Chapter Wrap-Up

Now that you have read the chapter, think about what you have learned. Complete the **What I Learned** *column on the first page of the chapter.*

Use this checklist to help you study.

- ❑ Complete your Foldables® Chapter Project.
- ❑ Study your *Science Notebook* on this chapter.
- ❑ Study the definitions of vocabulary words.
- ❑ Reread the chapter, and review the charts, graphs, and illustrations.
- ❑ Review the Understanding Key Concepts at the end of each lesson.
- ❑ Look over the Chapter Review at the end of the chapter.

THE BIG IDEA

Summarize It Reread the chapter Big Idea and the lesson Key Concepts. Analyze the information you have learned about DNA and genetics. How do genes, environment, and life choices affect a human's phenotype?

Challenge *Explain why DNA is vitally important to the cloning process.*

Name ____________________ Date ____________

The Environment and Change Over Time

How do species adapt to changing environments over time?

Before You Read

Before you read the chapter, think about what you know about how species change over time. Record three things that you already know about adaptation to environmental change in the first column. Then write three things that you would like to learn about change over time in the second column. Complete the final column of the chart when you have finished this chapter.

K What I Know	W What I Want to Learn	L What I Learned

Chapter Vocabulary

Lesson 1	Lesson 2	Lesson 3
NEW fossil record mold cast trace fossil geologic time scale extinction biological evolution **REVIEW** isotopes	**NEW** naturalist variation natural selection adaptation camouflage mimicry selective breeding **ACADEMIC** convince **REVIEW** biochemistry	**NEW** comparative anatomy homologous structure analogous structure vestigial structure embryology

Lesson 1 Fossil Evidence of Evolution

Scan *Lesson 1. Read the lesson titles and bold words. Look at the pictures. Identify three facts you discovered about fossils. Record your facts in your Science Journal.*

Main Idea	Details
The Fossil Record *I found this on page ______.*	**Characterize** *the* fossil record.

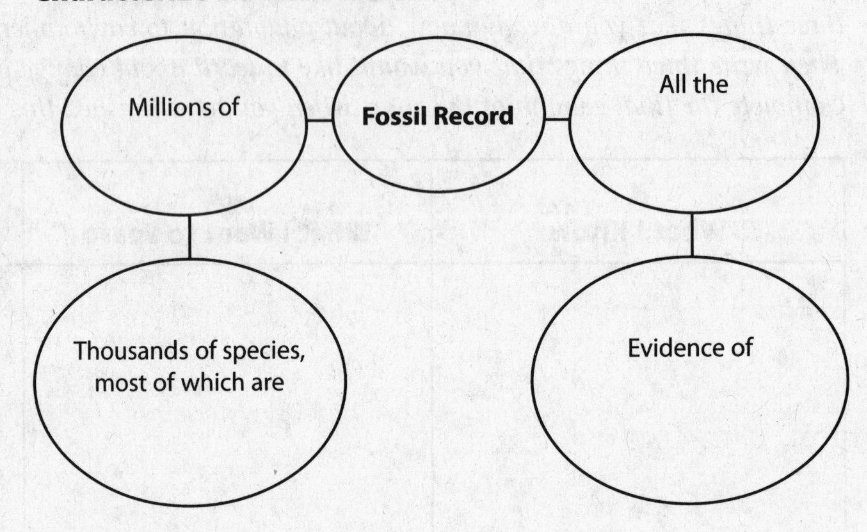

Fossil Formation

Examine *how fossils form.*

Main Idea	Formation	Explanation of the Process
I found this on page ______.	Mineralization	
I found this on page ______.	Carbonization	
I found this on page ______.	Molds and casts	
I found this on page ______.	Trace fossils	
I found this on page ______.	Original material	

Main Idea	Details

Determining a Fossil's Age

I found this on page ________.

Contrast *relative-age dating with absolute-age dating.*

________ Dating	________ Dating
• less precise • scientists compare • shows relative order in which	• more precise • scientists use • best measured in

Fossils over Time

I found this on page ________.

Order *the divisions of the* geologic time scale.

Eon	Era	Period	MYA

I found this on page ________.

Relate *types of fossils found in different rock layers to the* geologic time scale.

____

____

____

____

Main Idea — Details

Extinctions

I found this on page ______.

Relate *changes in the* fossil record *to mass* extinctions.

Fossil Record Changes	→ evidence →	Mass Extinctions

I found this on page ______.

Diagram *causes and effects of environmental change.*

Organisms depend on the environment for food and shelter. → Animals can't find resources they need to survive. → Result:

Quickly
Example:

Gradually
Example:

I found this on page ______.

Evaluate *how fossils provide scientists with evidence for* biological evolution.

______

______

Synthesize It Summarize how the fossil record provides evidence of both extinctions and biological evolution. Explain why it is important to look at fossils from multiple rock layers at the same time to reach conclusions.

______

______

______

______

______

Lesson 2 Theory of Evolution by Natural Selection

Predict *three facts that will be discussed in Lesson 2 after reading the headings. Write your facts in your Science Journal.*

Main Idea — Details

Charles Darwin

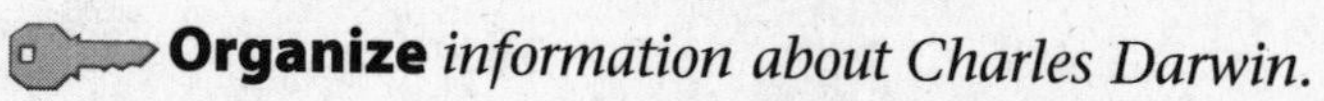

Organize *information about Charles Darwin.*

Main Idea		Charles Darwin
I found this on page ________.	Who?	
I found this on page ________.	What?	
I found this on page ________.	When?	
I found this on page ________.	Where?	
I found this on page ________.	Why?	
I found this on page ________.	How?	

I found this on page ________.

Correlate *species of tortoises with their food sources.*

Domed Tortoise	Intermediate Tortoise	Saddleback Tortoise

Darwin's Theory

I found this on page ________.

Explain *Darwin's theory of the relationship among the tortoises mentioned above.*

____

____

____

____

____

____

Main Idea	Details
I found this on page ________.	**Model** variation. *Draw several members of a species, and then explain how your drawing illustrates* variation.

Explanation: ____________________

I found this on page ________.

Describe *the process of* natural selection *that would result in the evolution of tortoises shown in picture 1 to those shown in picture 2.*

Picture 1

Picture 2

Main Idea	Details
Adaptations *I found this on page* ______.	**Relate** adaptations *to* natural selection. ______________________ ______________________ ______________________
I found this on page ______.	**Classify** *types of* adaptations.
I found this on page ______.	**Compare and contrast** camouflage *and* mimicry.
Artificial Selection *I found this on page* ______.	**Identify** *the process of breeding organisms for desired traits.* ______________, also known as ______________

Adaptations

Type:	Type:	Type:
Examples:	Examples:	Examples:

Camouflage	Both	Mimicry
•	•	•
	•	•

Connect It Explain how physical variation is expressed in human beings and how it relates to cultural groups and ethnicity.

____

____

____

____

____

Lesson 3 Biological Evidence of Evolution

Skim *Lesson 3 in your book. Read the headings and look at the photos and illustrations. Identify three things you want to learn more about as you read the lesson. Record your ideas in your Science Journal.*

Main Idea	Details
Evidence for Evolution *I found this on page* ______.	**Clarify** *why a diagram of evolution looks more like a bush than a straight line.* ______ ______ ______ ______ ______
I found this on page ______.	**Organize** *information about* comparative anatomy.
I found this on page ______.	**Comparative Anatomy** Definition:
I found this on page ______.	Homologous Structures: / Analogous Structures: / Vestigial Structures:
I found this on page ______.	
I found this on page ______.	**Sequence** *the development of the pharyngeal pouch in different species. Express the conclusion of scientists in* embryology.

All vertebrate embryos have a pharyngeal pouch. → **Development:** fish → ______ reptile → ______ bird → ______ human → ______ → **Conclusion:**

Main Idea | Details

I found this on page ________.

Identify *two ways in which molecular biology has affected the theory of evolution.*

1. ______________________________

2. ______________________________

I found this on page ________.

Assess *how genes support the theory of evolution.*

All organisms on Earth

All genes are

All organisms are

All genes work

Organisms came from

The Study of Evolution Today

I found this on page ________.

Relate *four areas of study that provide evidence of relationships between living and extinct species.*

Evidence for Evolution

Analyze It Use the analogy of building blocks to explain how genes relate to the diversity of life.

____

____

____

____

____

Review

The Environment and Change Over Time

Chapter Wrap-Up

Now that you have read the chapter, think about what you have learned. Complete the final column in the chart on the first page of this chapter.

Use this checklist to help you study.

- ❑ Complete your Foldables® Chapter Project.
- ❑ Study your *Science Notebook* on this chapter.
- ❑ Study the definitions of vocabulary words.
- ❑ Reread the chapter, and review the charts, graphs, and illustrations.
- ❑ Review the Understanding Key Concepts at the end of each lesson.
- ❑ Look over the Chapter Review at the end of the chapter.

THE BIG IDEA

Summarize It Reread the chapter Big Idea and the lesson Key Concepts. Summarize what you learned about species of horses throughout the chapter. Use as many of the chapter vocabulary words as you can.

Challenge *Choose an animal species that interests you. Research what scientists have discovered about the origin of that species. Draw a diagram that shows a "tree" with the animal's related species over evolutionary time. Share your diagram with your class.*

Name ________________ Date ________

Human Body Systems

What are the functions of the human body systems?

Before You Read

Before you read the chapter, think about what you know about the systems of the human body and their functions. Record your ideas in the first column. Pair with a partner, and discuss his or her thoughts. Write those thoughts in the second column. Then record what you both would like to share with the class in the third column.

Think	Pair	Share

Chapter Vocabulary

Lesson 1	Lesson 2	Lesson 3
NEW organ system homeostasis nutrient Calorie lymphocyte immunity **REVIEW** protein **ACADEMIC** detect	**NEW** compact bone spongy bone neuron reflex hormone	**NEW** reproduction gamete sperm ovum fertilization zygote

Transport and Defense

Skim *Lesson 1 in your book. Read the headings and look at the photos and illustrations. Identify three things you want to learn more about as you read the lesson. Record your ideas in your Science Journal.*

Main Idea	Details
The Body's Organization *I found this on page* ______.	**Relate** *the parts of an organism to each other.* Groups of cells that function together form ______. Groups of ______ that function together form ______. Groups of ______ that function together form ______. Organ systems ______ and maintain ______.
Digestion and Excretion *I found this on page* ______.	**Sequence** *the three major steps of the digestive process.* ______ → ______ → ______

Order *information about the digestive system. number.*

Main Idea	Order	Part	What happens there?
I found this on page ______.		small intestine	
I found this on page ______.		esophagus	
I found this on page ______.		large intestine	
I found this on page ______.	6	rectum	
I found this on page ______.	1	mouth	
I found this on page ______.		pancreas	

Main Idea	Details

I found this on page ______.

Compare and contrast *the pancreas and the liver.*

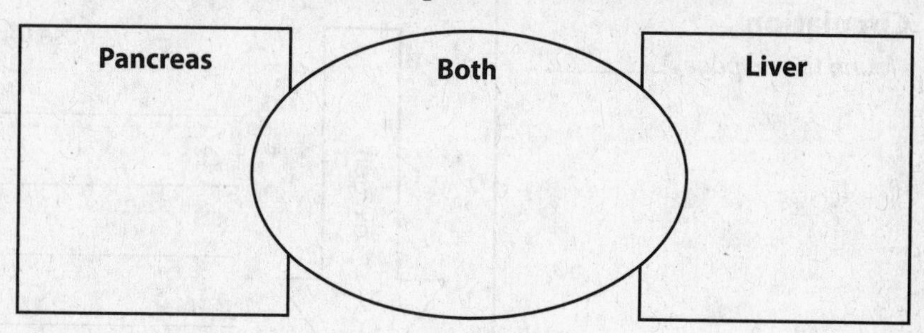

I found this on page ______.

Organize *information about nutrition.*

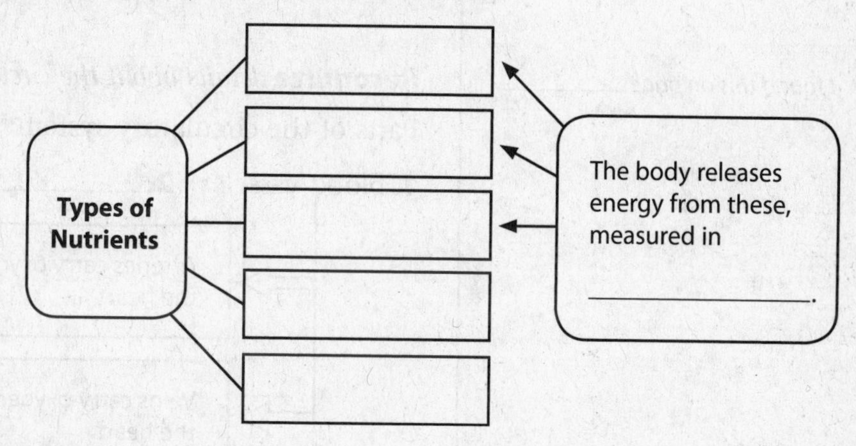

I found this on page ______.

Describe *the role of each part of the excretory system.*

Part	Role
lungs	
	removes salt and water when you sweat
rectum	

Main Idea	Details

Respiration and Circulation

I found this on page ________.

Sequence *the route that gases follow during respiration.*

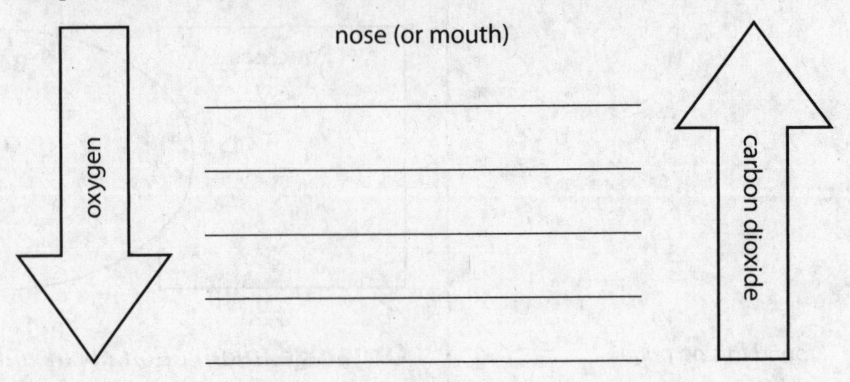

I found this on page ________.

Recognize *details about the circulatory system.*

Parts of the circulatory system:

1. blood vessels **2.** ________ **3.** ________

- Arteries carry oxygen-rich blood ________ the heart.
- Veins carry oxygen-poor blood ________ the heart.
- Capillaries enable substances to move ________ the circulatory system and the entire body.

I found this on page ________.

Characterize *the parts of blood.*

Part	Details
	the liquid part of blood; contains nutrients and allows transport of other blood cells
Platelets	

Main Idea | Details

I found this on page ______.

Identify *the parts and functions of the lymphatic system.*

Six Parts	Three Functions
• • • • • • lymphocytes	1. 2. 3.

I found this on page ______.

Define immunity.

____

I found this on page ______.

Distinguish *infectious from noninfectious diseases.*

	Infectious	Noninfectious
Cause		

I found this on page ______.

Describe *the body's lines of defense.*

First	Second	Third

Connect It Explain how the digestive, excretory, respiratory, and lymphatic systems all rely on the circulatory system.

____

____

____

____

Lesson 2 Structure, Movement, and Control

Scan *Lesson 2. Read the lesson titles and bold words. Look at the pictures. Identify three facts you discovered about the structure, movement, and control of the human body. Record your facts in your Science Journal.*

Main Idea	Details
Structure and Movement *I found this on page* ________.	**Identify** *the body systems that contribute to structure and movement.* **1.** ____________ **2.** ____________
I found this on page ________.	**Classify** *details about the skeletal system.*

Four Jobs	Parts	Bone Types
• • • •	• • • •	• •

Main Idea	Details
I found this on page ________.	**Differentiate** *the types of muscle tissue.*

Skeletal	Cardiac	Smooth

Main Idea	Details
Control and Coordination *I found this on page* ________.	**Identify** *the 2 body systems that receive and process information about internal and external environments.* **1.** ____________ **2.** ____________
I found this on page ________.	**Sequence** *the functioning of the nervous system.* **1.** ________ information **2.** ________ information **3.** ________ information

Main Idea	Details

I found this on page ______.

Categorize *parts of the nervous system.*

Central Nervous System:	Peripheral Nervous System:
	made of

I found this on page ______.

Differentiate *voluntary and involuntary functions of the nervous system by writing examples of each.*

Voluntary	Involuntary
•	•
•	•
•	• reflex

I found this on page ______.

Identify *the 5 ways in which humans detect their external environment.*

1. ______ **2.** ______ **3.** ______

4. ______ **5.** ______

I found this on page ______.

Relate *examples of how the body can respond to changes in its environment.*

	Change	Response
External	contact with a hot surface	
External	the smell of cookies	
Internal	release of the hormone insulin	
Internal	release of parathyroid hormone	

Analyze It Compare and contrast the nervous system with the endocrine system.

____

____

____

____

____

Lesson 3 Reproduction and Development

Predict *three facts that will be discussed in Lesson 3 after reading the headings. Record your predictions in your Science Journal.*

Main Idea / Details

Reproduction and Hormones

I found this on page ________.

Organize *information about* reproduction.

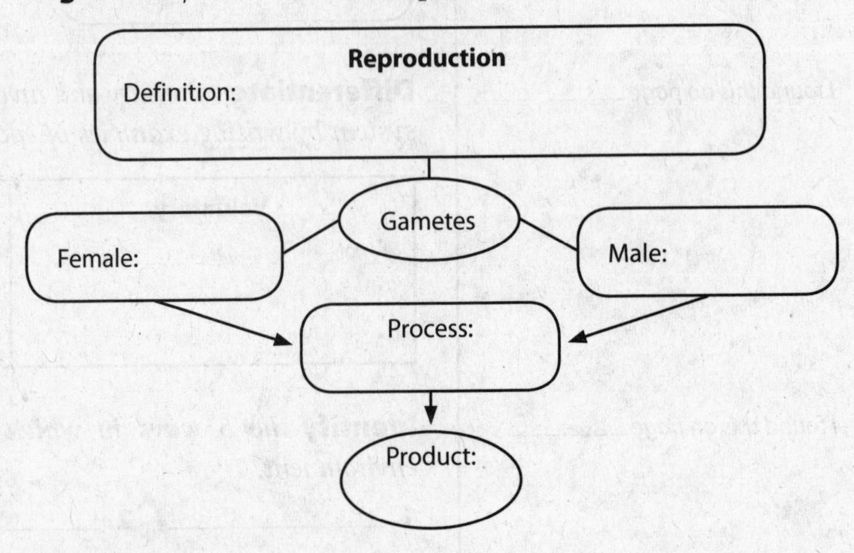

I found this on page ________.

Characterize *elements of the human reproductive system. Write* M *for* male *and* F *for* female.

M/F	Element	Description
	sperm	
	ovary	
	egg	
	scrotum	
	estrogen	
	semen	
	fallopian tubes	
	uterus	
	testis	
	progesterone	
	sperm duct	
	menstrual cycle	
	testosterone	
	penis	
	vagina	

Main Idea | **Details**

Human Development

I found this on page ______.

Sequence *development from* fertilization *through birth.*

A sperm fertilizes an egg and forms a zygote.

⇩

______

⇩

______

⇩

______

⇩

The fetus is pushed from the body through the vagina during birth.

I found this on page ______.

Compare *the stages of human development from birth through adulthood.*

Stage	When?	Major Characteristics
	birth–2 years	
	about 2–12 years	
	during puberty	
	after puberty	continued change
Aging	later adulthood	

Synthesize It Describe the stage of human development during which you believe the most dramatic changes occur. Explain your choice.

______

______

______

______

Review

Human Body Systems

Chapter Wrap-Up

Now that you have read the chapter, think about what you have learned.

Use this checklist to help you study.

- ❑ Complete your Foldables® Chapter Project.
- ❑ Study your *Science Notebook* on this chapter.
- ❑ Study the definitions of vocabulary words.
- ❑ Reread the chapter, and review the charts, graphs, and illustrations.
- ❑ Review the Understanding Key Concepts at the end of each lesson.
- ❑ Look over the Chapter Review at the end of the chapter.

THE BIG IDEA

Summarize It Reread the chapter Big Idea and the lesson Key Concepts. In a complete paragraph, write how each of the organ systems you read about in the chapter works with another system.

____

____

____

____

____

____

____

____

____

____

____

____

____

____

Challenge *Research how robots are designed, built, and maintained. Write a summary of how robotic systems compare and contrast with the organ systems of the human body. Illustrate your summary, and share it with your class.*

Name ______________________________ Date ______________

Plant Processes and Reproduction

What processes enable plants to survive and reproduce?

Before You Read

Before you read the chapter, think about what you know about plants and how they reproduce. Record your thoughts in the first column. Pair with a partner, and discuss his or her thoughts. Write those thoughts in the second column. Then record what you both would like to share with the class in the third column.

Think	Pair	Share

Chapter Vocabulary

Lesson 1	Lesson 2	Lesson 3
NEW photosynthesis cellular respiration **ACADEMIC** energy **REVIEW** molecule	**NEW** stimulus tropism photoperiodism plant hormone	**NEW** alternation of generations spore pollen grain pollination ovule embryo seed stamen pistil ovary fruit **REVIEW** mitosis

Lesson 1 Energy Processing in Plants

Scan *Lesson 1. Read the lesson titles and bold words. Look at the pictures. Identify three facts you discovered about energy processing in plants. Record your facts in your Science Journal.*

Main Idea	Details
Materials for Plant Processes *I found this on page ______.*	**Explain** *three functions that plants must be able to perform in order to survive.*

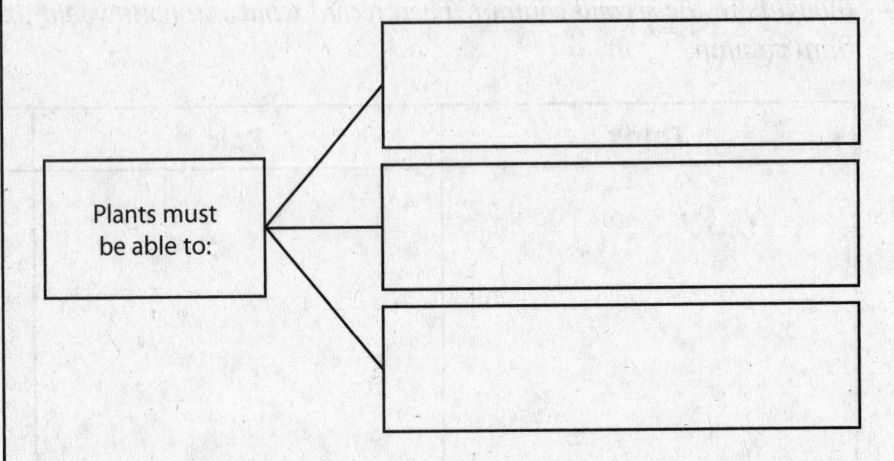

I found this on page ______.

Sequence *the movement of water inside plants.*

Water is absorbed by ______________________.

It travels inside ______________________,

up through ______________________,

and into ______________________.

I found this on page ______.

Review *the events that lead to a plant's wilting.*

lack of ____________ → ____________ collapse → plant wilts

Main Idea	Details

I found this on page ________.

Relate *how other materials move through plants.*

______________ produced by leaves	Gases such as ______________ ______________, ______________, and ______________
moves	move
throughout the plant through the ______________.	into and out of the plant through ______________ ______________.

Photosynthesis
I found this on page ________.

Order *the chemical reactions of* photosynthesis.

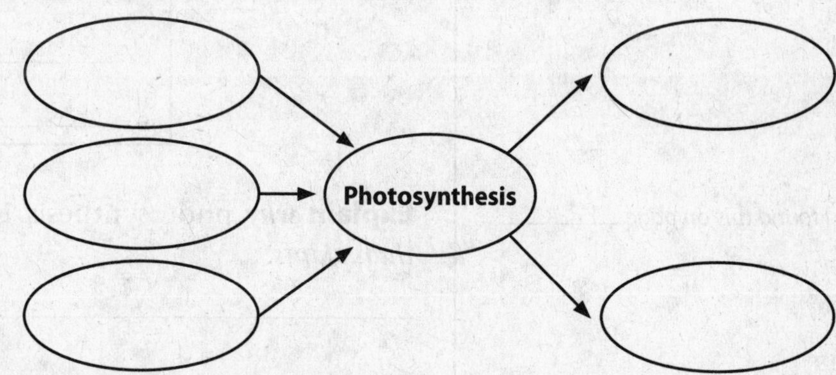

I found this on page ________.

Organize *information about the cells of leaves.*

Part	Description and Function
	flat, irregularly shaped cells that make up the top and bottom layers of leaves
Stomata	______________ on the lower epidermal layer that ______________ move through
	internal cells packed together to expose the greatest number to ______________
Chloroplast	the ______________ inside the ______________, where photosynthesis occurs

Main Idea	Details
I found this on page ______.	**Sequence** *the process of* photosynthesis.

1. ______ in the chloroplast captures ______.
2. ______ is transferred to other molecules.
3. ______ are split apart; ______ is released.
4. ______ is converted into ______ and combined with ______ ______ to form ______.

I found this on page ______.

Explain *why* photosynthesis *is important to organisms other than plants.*

______

______

______

______

Cellular Respiration

I found this on page ______.

Complete *the diagram of* cellular respiration.

( ) → cellular respiration → ( )

Main Idea	Details
I found this on page ________.	**Diagram** *the chemical reactions of* cellular respiration.

Occurs in the mitochondria and cytoplasm

Main Idea	Details
I found this on page ________.	**Connect** photosynthesis *and* cellular respiration *with their characteristics. Draw a line from each term to all of the descriptors that apply to it.*

Cellular Respiration •

Photosynthesis •

- makes food energy
- performed in plants and algae
- converts food energy
- requires sunlight, carbon dioxide, and water
- occurs in chloroplast
- produces carbon dioxide, water, and ATP
- occurs in mitochondria
- produces glucose and oxygen
- performed in all organisms
- requires glucose and oxygen

Synthesize It Photosynthesis and cellular respiration are both chemical reactions that occur inside cells. Explain, in your own words, how the two processes are alike and different.

____

____

____

____

Lesson 2 Plant Responses

Predict *three facts that will be discussed in Lesson 2 after reading the headings. Write your facts in your Science Journal.*

Main Idea — Details

Stimuli and Plant Responses
I found this on page ________.

Identify *examples of plant* stimuli *and* responses.

	Stimulus	Response
Gradual		
Quick		

Environmental Stimuli
I found this on page ________.

Characterize tropisms. *Tell whether each response is positive or negative.*

Tropism	Stimulus	Response (positive or negative)	
	light	stem:	roots:
	touch	drooping:	tendrils:
Gravitropism		stems:	roots:

I found this on page ________.

Review *and give an example of each type of* photoperiodism.

1. Long-Day Plants: ________________

____

2. Short-Day Plants: ________________

____

3. Day-Neutral Plants: ________________

____

Main Idea | Details

Chemical Stimuli and **Summary of Plant Hormones**

 **Relate** *the effects of* plant hormones.

Plant Hormone	Chemical Message(s)
Auxins	
Ethylene	
Gibberellins	
Cytokynins	

I found this on page ______.

I found this on page ______.

I found this on page ______.

I found this on page ______.

Humans and Plant Responses

I found this on page ______.

Identify *four needs for which humans depend on plants.*

1. ______ **3.** ______

2. ______ **4.** ______

I found this on page ______.

Explain *two ways that people can benefit from understanding and using plant responses. Give examples.*

Plant Response	Example of Use or Benefit
Climbing vines	
Ripening in response to ethylene	

Connect It Explain why it would be financially beneficial for a farmer to treat a fruit crop with gibberellins.

______

______

______

______

______

Lesson 3 Plant Reproduction

Skim *Lesson 3 in your book. Read the headings and look at the photos and illustrations. Identify three things you want to learn more about as you read the lesson. Record your ideas in your Science Journal.*

Main Idea	Details

Asexual Reproduction Versus Sexual Reproduction

I found this on page ______.

Differentiate *asexual reproduction and sexual reproduction in plants.*

Asexual Reproduction	Sexual Reproduction
What develops into a new plant?	What develops into a new plant?
How does it compare genetically to the parent plant?	How does it compare genetically to the parent plants?

Alternation of Generations

I found this on page ______.

Model *the* alternation of generations *in plants.*

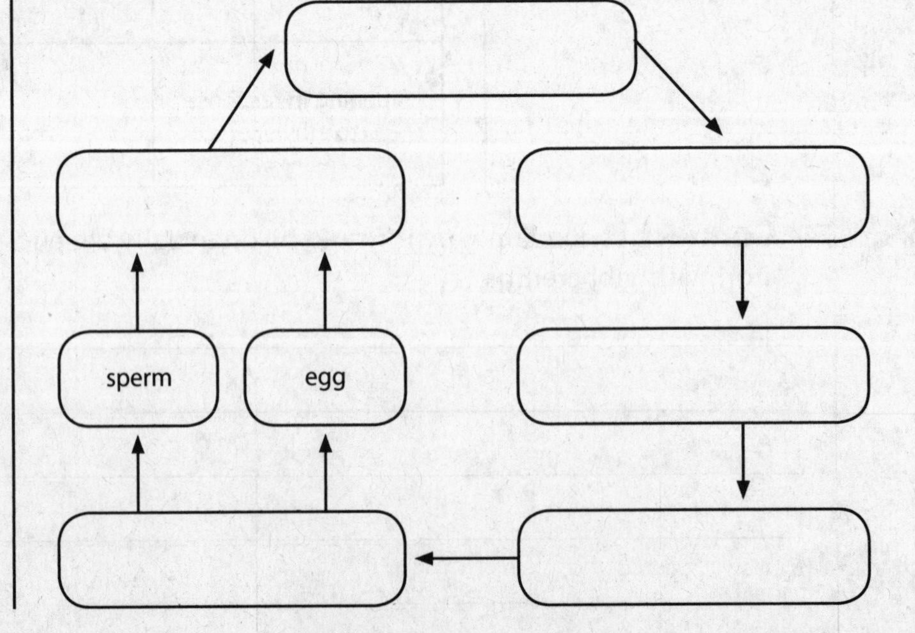

Main Idea	Details

Reproduction in Seedless Plants

I found this on page ______.

Order *the stages in the life cycles of moss and ferns. Draw an arrow from the last step in the cycle back to the step numbered* 1.

	Life Cycle of Moss and Ferns
	Zygote grows by mitosis into diploid generation plant.
	Haploid spores grow into tiny plants that have male structures and female structures.
1	Diploid generation produces haploid spores.
	Male structures produce sperm and female structures produce eggs.
	Fertilization results in a diploid zygote.

How do seed plants reproduce?

I found this on page ______.

Identify *the 2 types of* seed *plants.*

1. ______

2. ______

I found this on page ______.

Explain *the relationship between* pollen grains, ovules, *and* seeds. *Use the terms* pollination *and* embryo *in your explanation.*

______

______

______

______

______

Main Idea | **Details**

I found this on page ______.

Draw *a cross section of a seed that shows its three main parts. Explain the role of each part.*

	Covering
	Food Supply
	Embryo

I found this on page ______.

Define *gymnosperm, and tell why certain plants have that name.*

I found this on page ______.

Model *the life cycle of a gymnosperm.*

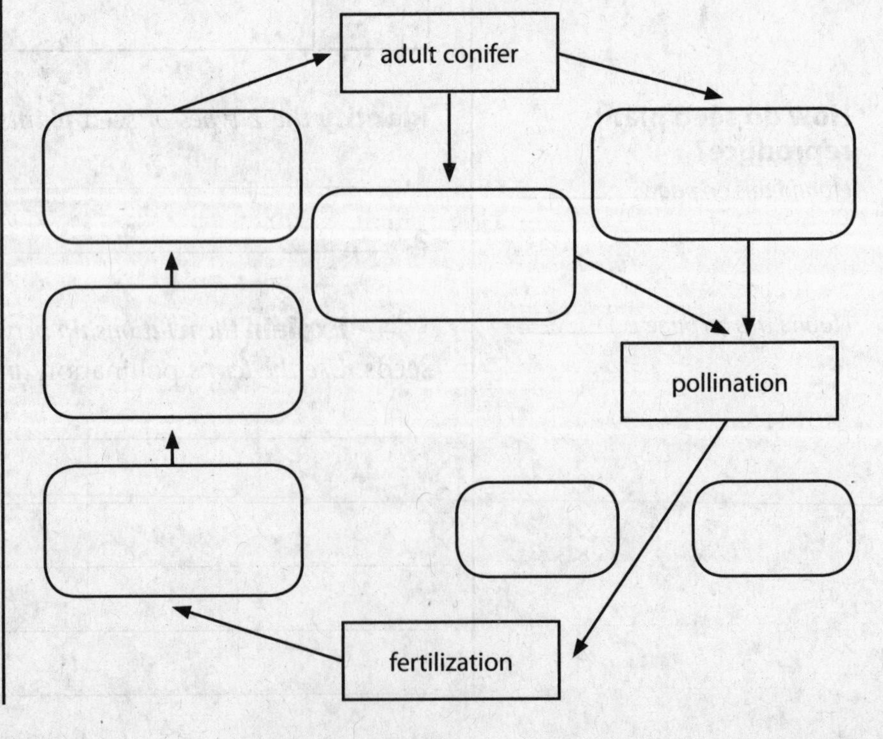

Main Idea	Details

I found this on page ________.

Identify *the parts of a flower.*

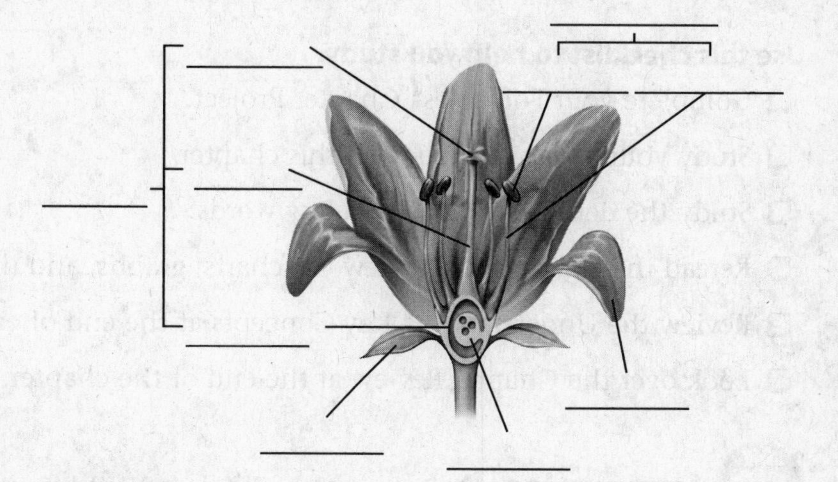

I found this on page ________.

Summarize *the life cycle of an angiosperm.*

Structure	What It Does
Mature diploid plant	produces flowers
Pollen	
Sperm and ovule	
Zygote	
Ovary and ovule	
Seed	

I found this on page ________.

Relate *how* fruit *can promote* seed *dispersal.*

Animal eats fruit. → [ ] → New plant sprouts in different location.

Analyze It What advantage do diverse methods of reproduction provide plants? Give examples.

____

____

____

Review

Plant Processes and Reproduction

Chapter Wrap-Up

Now that you have read the chapter, think about what you have learned.

Use this checklist to help you study.

- ❑ Complete your Foldables® Chapter Project.
- ❑ Study your *Science Notebook* on this chapter.
- ❑ Study the definitions of vocabulary words.
- ❑ Reread the chapter, and review the charts, graphs, and illustrations.
- ❑ Review the Understanding Key Concepts at the end of each lesson.
- ❑ Look over the Chapter Review at the end of the chapter.

THE BIG IDEA

Summarize It Reread the chapter Big Idea and the lesson Key Concepts. Summarize how plants are the key to humans surviving the Sun's energy. Connect concepts from all three lessons in your answer.

Challenge *Grow a potted plant indoors at home, and choose an outdoor plant to observe. Observe both plants over an extended time period of a couple months. Keep a journal comparing the processes, stimuli, and responses of the plants. Summarize your observations and share the comparison with your class.*

Name ______________________ Date __________

Interactions of Living Things

How do living things interact with and depend on the other parts of an ecosystem?

Before You Read

Before you read the chapter, think about what you know about the interactions of living things. Record your thoughts in the first column. Pair with a partner, and discuss his or her thoughts Record those thoughts in the second column. Then record what you both would like to share with the class in the third column.

Think	Pair	Share

Chapter Vocabulary

Lesson 1	Lesson 2	Lesson 3
NEW	**NEW**	**NEW**
ecosystem	limiting factor	producer
abiotic factor	biotic potential	consumer
biotic factor	carrying capacity	food chain
population	habitat	food web
community	niche	energy pyramid
biome	symbiotic relationship	
succession		**ACADEMIC**
		source
		transform

Lesson 1 Ecosystems and Biomes

Scan *Lesson 1. Then write three questions that you have about ecosystems and biomes in your Science Journal. Try to answer your questions as you read.*

Main Idea	Details
What are ecosystems? *I found this on page* ______.	**Record** *the 2 parts that make up an* ecosystem. **1.** ______ **2.** ______
I found this on page ______.	**Draw** *an* ecosystem, *and label at least eight of its parts.*
Abiotic Factors	**Identify** *five important* abiotic factors *found on Earth, and describe the effect each factor has on an* ecosystem.

I found this on page ______.

I found this on page ______.

I found this on page ______.

I found this on page ______.

I found this on page ______.

Factor	Effect
Water	
	Nutrients, texture, and depth determine which organisms can survive and thrive.

Main Idea	Details
Biotic Factors *I found this on page* ________.	**Describe** biotic factors. ________________ ________________
I found this on page ________.	**Identify** *the role of individuals in a* population.

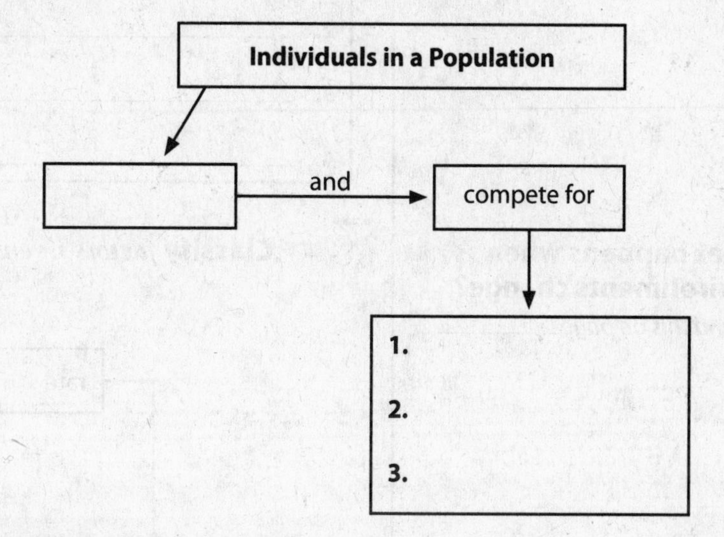

I found this on page ________.	**Relate** populations *to* communities. ________________ ________________ ________________
I found this on page ________.	**Identify** *the components of a* biome.

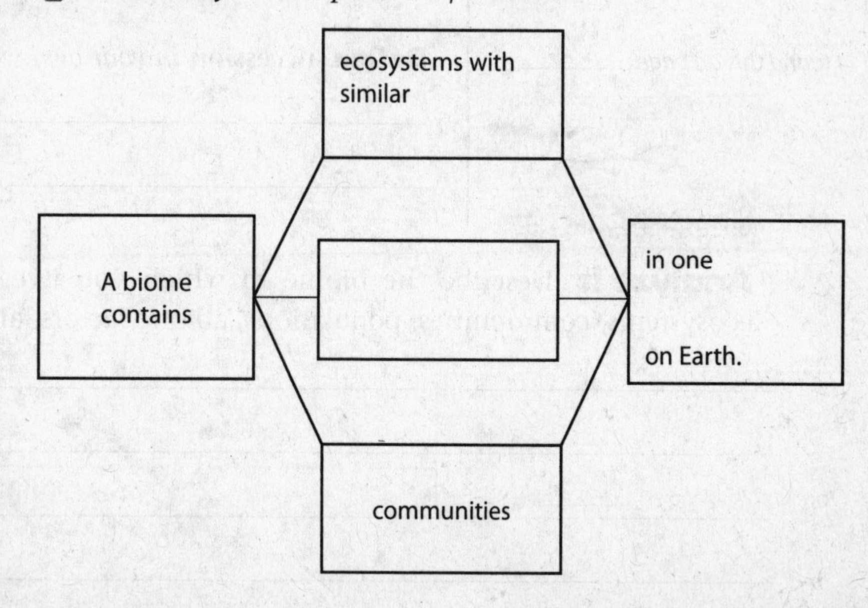

Main Idea / Details

I found this on page ________.

Categorize *Earth's* biomes *as terrestrial or aquatic.*

Biomes	
Terrestrial	**Aquatic**
1. ________	1. ________
2. ________	2. ________
3. ________	
4. ________	

What happens when environments change?

I found this on page ________.

Classify *factors in environmental change.*

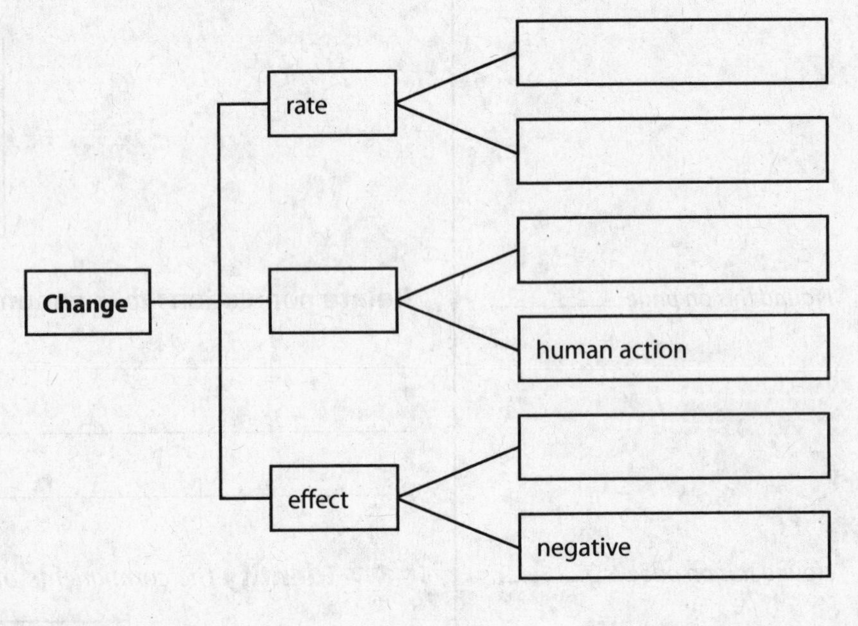

I found this on page ________.

Define succession *in your own words.*

____

____

Analyze It Describe the biome in which you live. Give multiple examples of ecosystems, communities, populations, abiotic factors, and biotic factors.

____

____

____

____

Lesson 2 Populations and Communities

Predict *three facts that will be discussed in Lesson 2 after reading the headings. Record your facts in your Science Journal.*

Main Idea — Details

Populations

I found this on page ______.

Categorize *causes of population changes.*

Change	Causes
Size increase	1. 2.
Size decrease	1. 2.

I found this on page ______.

Complete *the diagram below with arrows to characterize population density.*

Size of population [ ] Relative space available [ ]

I found this on page ______.

Define limiting factors, *and give four examples of these factors.*

limiting factors: ______________________

____

1. ______ **3.** ______

2. ______ **4.** ______

I found this on page ______.

Compare and contrast biotic potential *with* carrying capacity.

____

____

____

____

____

I found this on page ______.

Identify *three effects of overpopulation.*

Overpopulation → ______________________

Overpopulation → ______________________ environment

Overpopulation → lack of ______________________

Main Idea	Details

Communities
I found this on page ________.

Describe *interactions among populations in a lake community.*

Population	Interaction
Insects	
Plants	
Fish	
Ducks	
Turtles	

Symbiotic Relationships
I found this on page ________.

Contrast *a* habitat *with a* niche.

____

____

____

____

I found this on page ________.

Compare symbiotic relationships *by completing the table.*

	Mutualism	Parasitism	Commensalism
Number of species involved			
Number benefited			
Number harmed			
Number neither benefited nor harmed			

Synthesize It Use what you know about populations and communities to explain two possible outcomes of a growing human population.

____

____

____

Lesson 3 Energy and Matter

Skim *Lesson 3 in your book. Read the headings and look at the photos and illustrations. Identify three things you want to learn more about. Record your ideas in your Science Journal.*

Main Idea	Details

Energy Flow

I found this on page ________.

Draw *representations of a flow and a cycle.*

Flow	Cycle

Organisms and Energy

I found this on page ________.

Compare and contrast *the processes* producers *use to make food.*

Photosynthesis

converts ________

into ________

Both

use ________ in the environment

for ________

____

Chemosynthesis

converts ________

________ into ________

____

Assess *information about* consumers. *Read each statement. If it is true, write* **T** *in the center column. If it is false, write* **F** *in the center column and then rewrite the underlined words to make the statement true.*

Statement	T or F	Correction
Herbivores eat <u>consumers</u>.		
Omnivores eat <u>only detritivores</u>.		
<u>Carnivores</u> eat consumers.		
Detritivores <u>are predators</u>.		

I found this on page ________.

I found this on page ________.

I found this on page ________.

I found this on page ________.

Main Idea	Details

Modeling Energy Flow

I found this on page ________.

Draw *arrows to complete a* food web.

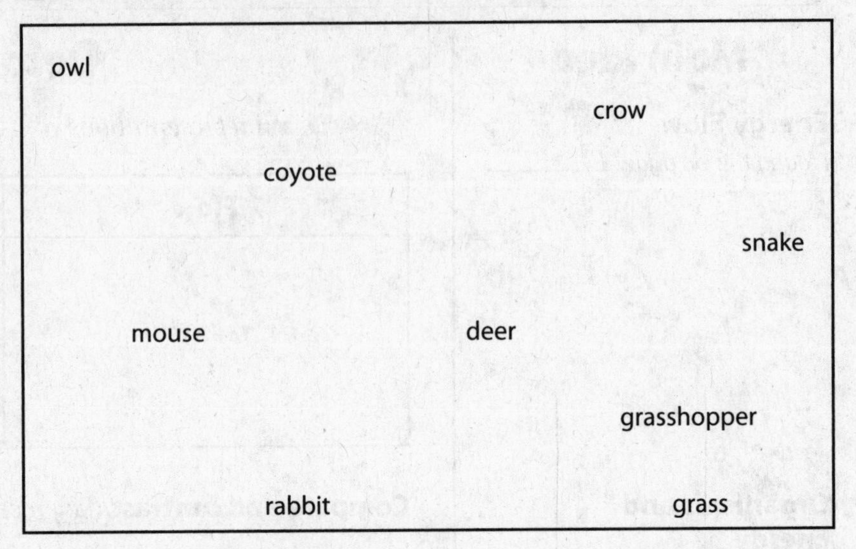

I found this on page ________.

Modify *the diagram above to show specific* food chains. *Use two different colors to trace the flow of energy through two* food chains.

Modeling Energy Pyramids

I found this on page ________.

Order *these organisms according to the amount of energy available. Place the organism with the most energy available at the bottom of the pyramid.*

algae   eagle   fish   tadpole

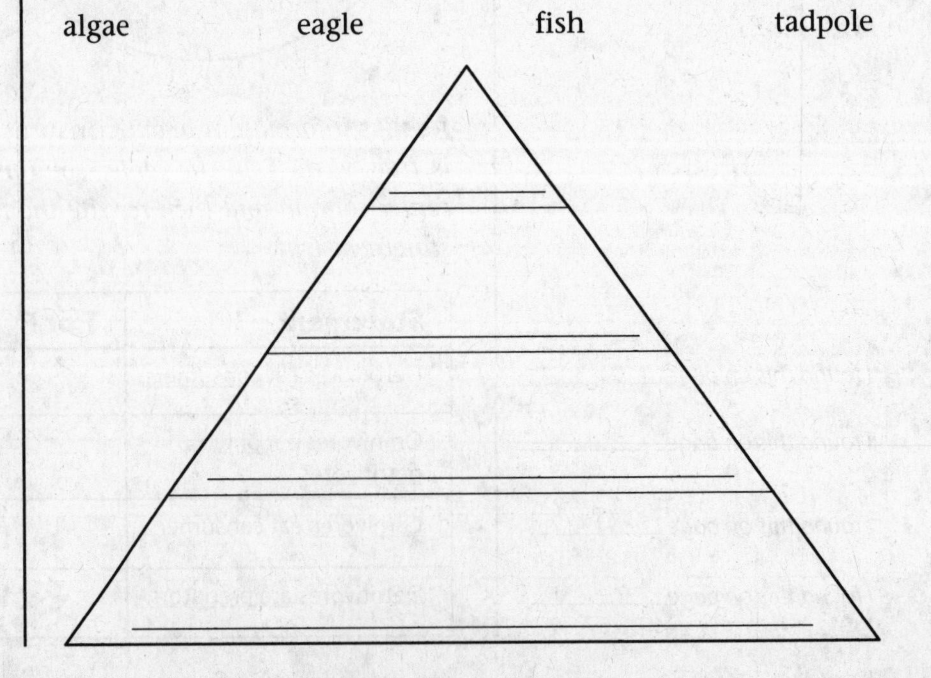

Main Idea | Details

Matter Cycles

I found this on page ________.

Model *the processes of the water cycle.*

evaporation: ________

____

________:

Water falls as ________

or ________.

I found this on page ________.

Explain *the oxygen cycle by placing arrows in the table.*

photosynthesis		carbon dioxide		respiration
		oxygen		

I found this on page ________.

Classify *processes in the carbon cycle.*

Removes carbon from the atmosphere	Releases carbon into the atmosphere	Releases carbon into the soil
	1.	
	2.	
	3.	

Analyze It Use what you know about energy and matter to describe at least three ways in which you affect and are affected by energy flow and the cycling of matter.

____

____

____

____

____

____

____

Review Interaction of Living Things

Chapter Wrap-Up

Now that you have read the chapter, think about what you have learned.

Use this checklist to help you study.

- ❑ Complete your Foldables® Chapter Project.
- ❑ Study your *Science Notebook* on this chapter.
- ❑ Study the definitions of vocabulary words.
- ❑ Reread the chapter, and review the charts, graphs, and illustrations.
- ❑ Review the Understanding Key Concepts at the end of each lesson.
- ❑ Look over the Chapter Review at the end of the chapter.

THE BIG IDEA **Summarize It** Reread the chapter Big Idea and the lesson Key Concepts. Draw models that show the interactions of key ideas from Lessons 1, 2, and 3. Label your drawing with key terms, and use arrows to show relationships. Describe your drawing in two or three sentences below.

Challenge *Science fiction stories often contain a great deal of factual information. Write a fictional story that describes factual challenges of living in an environment without Earth's natural matter cycles, such as what would be needed for humans to colonize another planet. Share your story with your class.*

Name ________________________________ Date ______________

Foundations of Chemistry

What is matter, and how does it change?

Before You Read

Before you read the chapter, think about what you know about matter and how it changes. Record three things that you already know about matter in the first column. Then write three things that you would like to learn about matter in the second column. Complete the final column of the chart when you have finished this chapter.

K What I Know	W What I Want to Learn	L What I Learned

Chapter Vocabulary

Lesson 1	Lesson 2	Lesson 3	Lesson 4
NEW matter atom substance element compound mixture heterogeneous mixture homogeneous mixture dissolve **ACADEMIC** unique	**NEW** physical property mass density solubility **REVIEW** property	**NEW** physical change	**NEW** chemical property chemical change concentration

Lesson 1 Classifying Matter

Scan *Lesson 1. Read the lesson titles and bold words. Look at the pictures. Identify three facts you discovered about matter. Record your facts in your Science Journal.*

Main Idea — Details

Understanding Matter
I found this on page ______.

Organize *information about* matter.

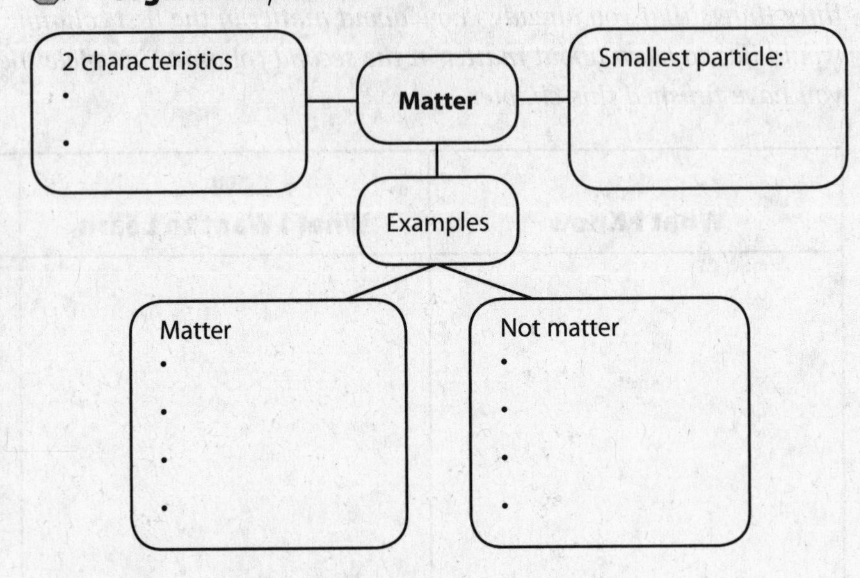

Atoms
I found this on page ______.

Identify and describe *the parts of an* atom.

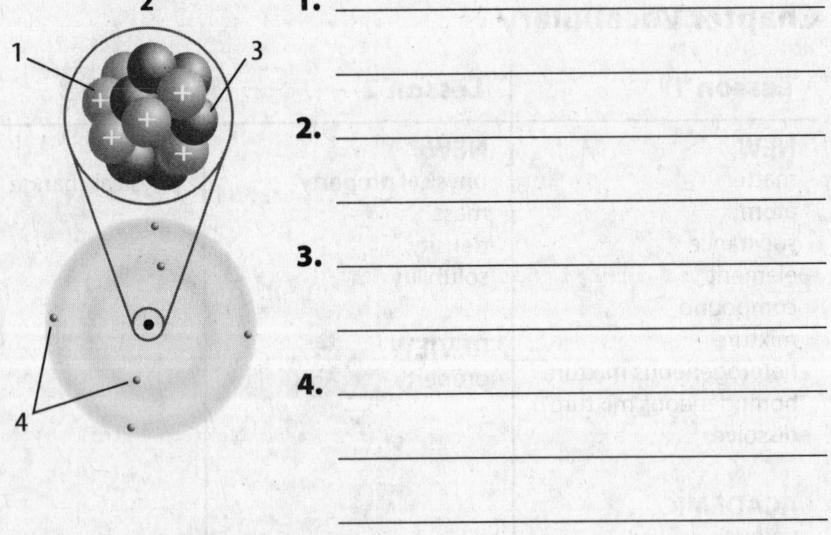

1. ______
2. ______
3. ______
4. ______

I found this on page ______.

Relate *the number of protons in the nucleus of an* atom *to the properties of* matter.

______

Main Idea	Details

Substances
I found this on page ______.

Define substance, *and identify two examples.*

Definition: ______________________________

____

Examples: ______________________________

I found this on page ______.

Examine *how* atoms *of* elements *differ. Circle the characteristic that gives each type of* atom *its unique properties.*

Example	Aluminum	Oxygen
Is it a substance?		
Is it an element?		
How many protons are in its nucleus?		
How do its atoms usually exist?		

I found this on page ______.

Differentiate elements *from* compounds.

Element	Compound

I found this on page ______.

Interpret *the chemical formula.*

This means

This means

CO_2

This means

I found this on page ______.

Relate *properties of a* compound *to the properties of the* elements *of which it is made.*

____

____

Main Idea	Details

Mixtures

I found this on page ________.

Organize *information about* mixtures.

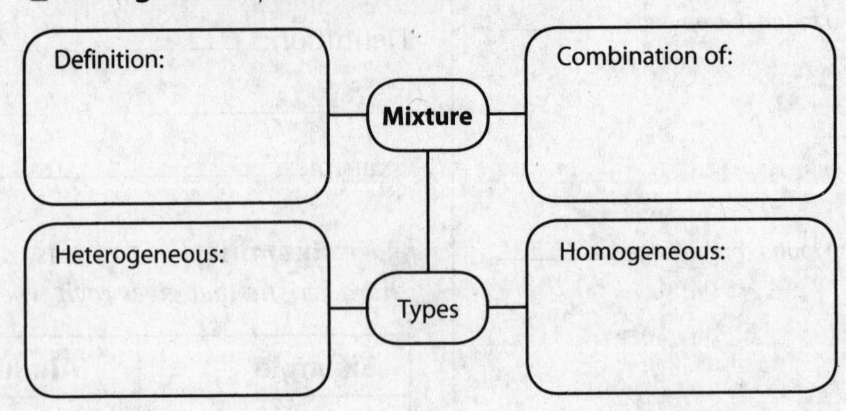

I found this on page ________.

Complete *the statement to describe how a* substance dissolves.

In a solution, the ________ is the ________ present in the largest amount; the ________ are all the other substances that ________ in the solvent.

Compounds v. Solutions

I found this on page ________.

Explain *whether a chemical formula can describe a solution.*

____

____

Summarizing Matter

I found this on page ________.

Order *the classifications of* matter. *Use the words in the box to make a sequence that is true.*

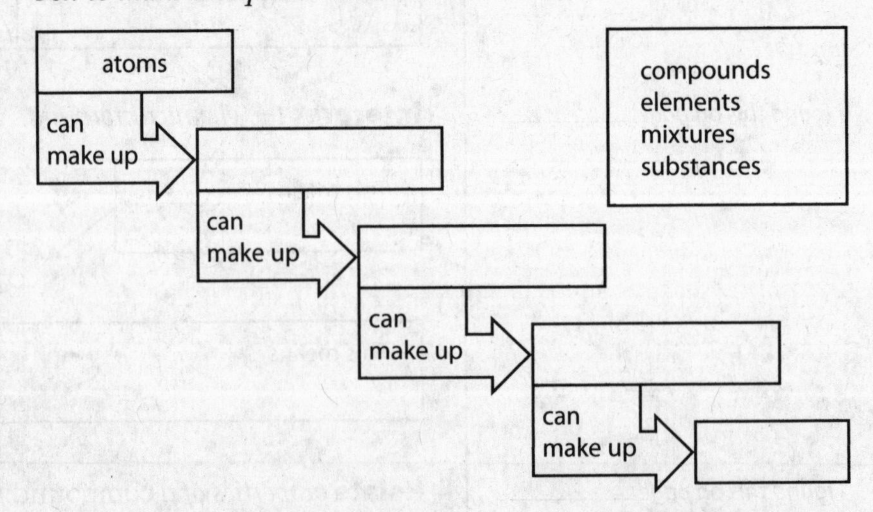

Analyze It Evaluate whether the following statement is true or false, and explain why: All solutions are mixtures.

____

Lesson 2 Physical Properties

Predict *three facts that will be discussed in Lesson 2 after reading the headings. Record your predictions in your Science Journal.*

Main Idea	Details

Physical Properties

I found this on page ________.

Define physical property, *and give two examples.*

Definition: ____________________

____

1. ____________ **2.** ____________

I found this on page ________.

Differentiate *states of matter.*

State	Arrangement of Particles	Motion of Particles
Solid		
Liquid		
Gas		

I found this on page ________.

Contrast *three size-dependent* physical properties. *Circle the measurement that changes with location.*

I found this on page ________.

Property	Description

Main Idea	Details

Describe *four size-independent* physical properties.

I found this on page ______.

Melting and boiling point:

Density:

Size-Independent Properties

I found this on page ______.

Conductivity:

Solubility:

I found this on page ______.

Identify and describe *three* physical properties *that can be used to separate mixtures.*

Property	How it can be used to separate a mixture

Synthesize It Look again at the picture of the man panning for gold on the first page of Lesson 2. Another method he could use to separate the mixture would be to sift the sediment through a screen. What physical property affects how sifting separates a mixture? Would this be as effective for finding gold as panning?

______

______

______

______

______

Lesson 3 Physical Changes

Skim *Lesson 3 in your book. Read the headings and look at the photos and illustrations. Identify three things you want to learn more about as you read the lesson. Record your ideas in your Science Journal.*

Main Idea	Details
Physical Changes *I found this on page* ________.	**Characterize** physical changes *in matter.* **Physical Changes in Matter** What can change? • • • • What does not change?
I found this on page ________.	**Explain** *one way that you changed the size or shape of matter as you made and ate your breakfast.* ________________ ________________
I found this on page ________.	**Sequence** *changes in the state of matter with the continuous addition of thermal energy.* Thermal energy is added to a solid; particles ________; temperature ________. ↓ Particles overcome attractive forces; ________ occurs; temperature ________. ↓ Entire solid becomes ________; particles move faster; temperature ________. ↓ Particles overcome attractive forces; ________ occurs; temperature ________. ↓ Entire liquid becomes ________; temperature ________.

Main Idea	Details
I found this on page ______.	**Identify** *the opposites of selected* physical changes.

melting →
boiling →
sublimation →

Main Idea	Details
I found this on page ______.	**Explain** *dissolving, and express how boiling can reverse the process in the example of salt water.*
Conservation of Mass *I found this on page ______.*	**Model** *conservation of mass in a labeled drawing that illustrates a mixture.*

Connect It Look at this paper in front of you and the pen or pencil in your hand. Describe the physical change and conservation of mass you observed in those items as you completed the exercises on this page.

Lesson 4 Chemical Properties and Changes

Predict *three facts that will be discussed in Lesson 4 after reading the headings. Record your predictions in your Science Journal.*

Main Idea / Details

Chemical Properties
I found this on page ________.

Define chemical property, *and give two examples.*

Definition: ____________________

____

1. ____________________

2. ____________________

Comparing Properties
I found this on page ________.

Contrast *physical and* chemical properties *using the example of a wood log.*

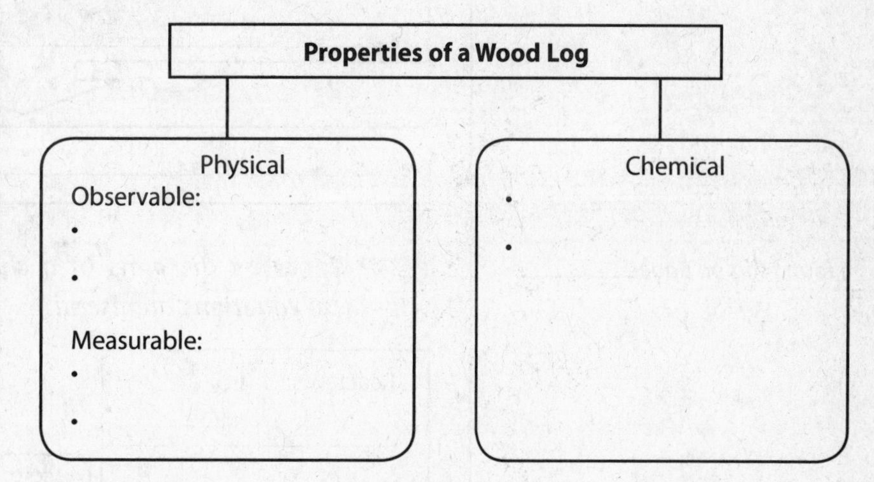

Chemical Changes
I found this on page ________.

Characterize chemical change.

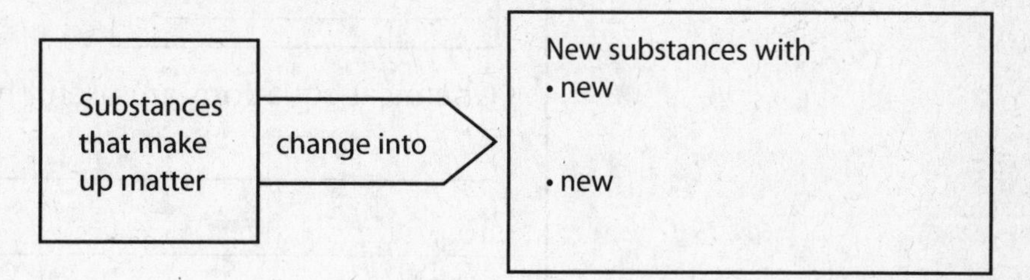

Signs of Chemical Change
I found this on page ________.

Identify *some signs of a* chemical change.

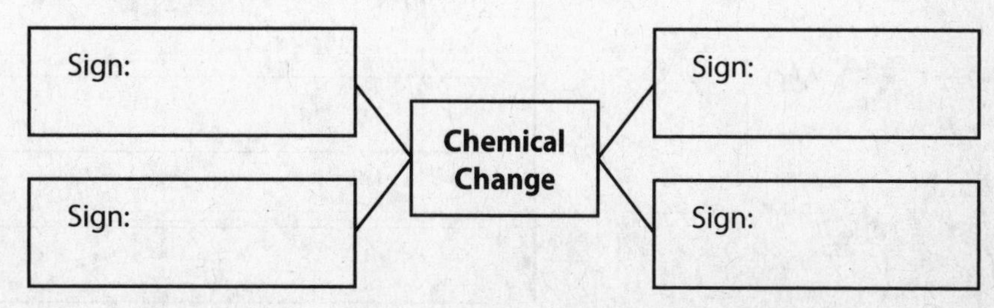

Main Idea	Details
I found this on page ______.	**Describe** *what constitutes proof of a* chemical change.
	______
Explaining Chemical Reactions I found this on page ______.	**Order** *the events that occur in a chemical reaction.*

Atoms are ______ together and form particles of substances.

↓

As particles collide, bonds break and atoms separate

↓

Atoms ______ ______.

↓

______ form.

I found this on page ______.

Describe *the parts of a chemical equation, and then tell why these equations are useful.*

Reactants:		Products:
______	yield →	______
______		______
______		______

Chemical equations are useful because ______

______

______

I found this on page ______.

Express *how a balanced chemical equation illustrates conservation of mass.*

______

______

______

______

Main Idea | Details

I found this on page ______.

Assess *the role of coefficients in chemical equations.*

Coefficients
- change
- do not change

The Rate of Chemical Reactions

I found this on page ______.

Explain *how factors affect the rate of chemical reactions.*

Factor	Effect on Reaction Speed
Temperature	
Concentration	
Surface area	

Chemistry

I found this on page ______.

Paraphrase *two things you need to know about matter in order to understand chemistry.*

1. ______

2. ______

Synthesize It Describe two chemical changes that have happened in your home this week.

______

Review

Foundations of Chemistry

Chapter Wrap-Up

Now that you have read the chapter, think about what you have learned. Complete the final column in the chart on the first page of the chapter.

Use this checklist to help you study.

- ❑ Complete your Foldables® Chapter Project.
- ❑ Study your *Science Notebook* on this chapter.
- ❑ Study the definitions of vocabulary words.
- ❑ Reread the chapter, and review the charts, graphs, and illustrations.
- ❑ Review the Understanding Key Concepts at the end of each lesson.
- ❑ Look over the Chapter Review at the end of the chapter.

THE BIG IDEA

Summarize It Reread the chapter Big Idea and the lesson Key Concepts. Summarize why it is necessary to have these ways of classifying and describing matter and its changes.

____

____

____

____

____

____

____

____

____

____

____

____

____

Challenge *Find evidence of a chemical reaction that has occurred at your home. Do research to learn about the matter involved—the reactants and the products. Write a balanced chemical equation for the reaction, and explain the equation to your class.*

Name _______________________________ Date _______________

The Periodic Table

How is the periodic table used to classify and provide information about all known elements?

Before You Read

Before you read the chapter, think about what you know about the periodic table. Write in the first column three things that you already know about how the periodic table is arranged and what it is used for. Then in the second column indicate three things that you would like to learn about the periodic table. When you have completed the chapter, think about what you have learned and complete the **What I Learned** *column.*

K What I Know	W What I Want to Learn	L What I Learned

Chapter Vocabulary

Lesson 1	Lesson 2	Lesson 3
NEW periodic table group period	**NEW** metal luster ductility malleability alkali metal alkaline earth metal transition element **REVIEW** density	**NEW** nonmetal halogen noble gas metalloid semiconductor **ACADEMIC** construct

Lesson 1 Using the Periodic Table

Skim *Lesson 1 in your book. Read the headings, and look at the photos and illustrations. Identify three things you want to learn more about as you read the lesson. Record your ideas in your Science Journal.*

Main Idea	Details
What is the periodic table? *I found this on page* ______.	**Define** periodic table. ______________________________ ______________________________ ______________________________
I found this on page ______.	**Organize** *information about the* periodic table *in the chart below.* There are more than ______ known elements. → Scientists use a ______ to organize the elements. → The periodic table is
Developing a Periodic Table *I found this on page* ______.	**Discuss** *information about Mendeleev.* **Dimitri Mendeleev** Who he was: What he developed:
I found this on page ______.	**Name** *four properties of elements that Mendeleev studied when developing his* periodic table. **1.** ______ **3.** ______ **2.** ______ **4.** ______

Main Idea	Details
I found this on page ______.	**Recall** *the definition of atomic number.* ______ ______ ______
I found this on page ______.	**Compare** *Mendeleev's* periodic table *with that of Moseley by completing the Venn diagram.*

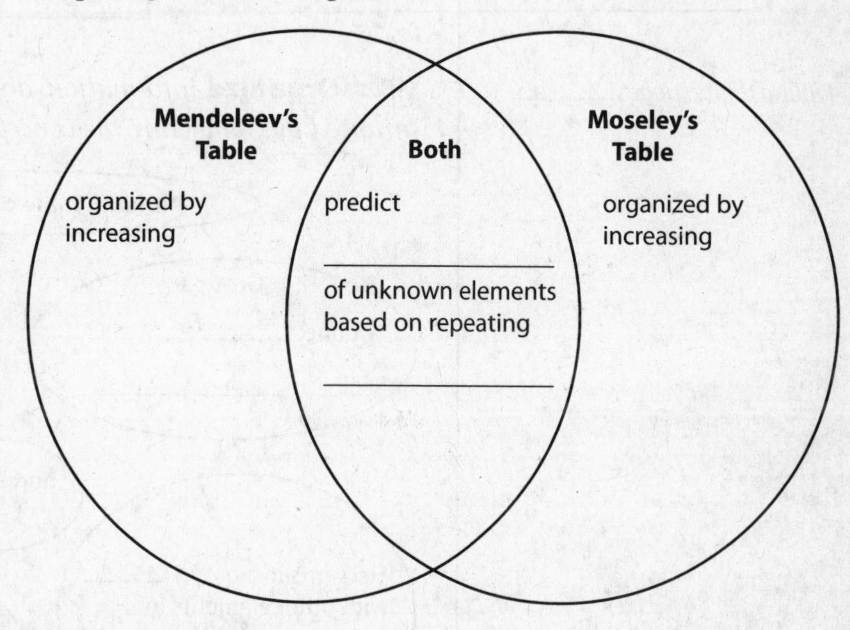

Today's Periodic Table
I found this on page ______.

Discuss *today's* periodic table *in the organizer below.*

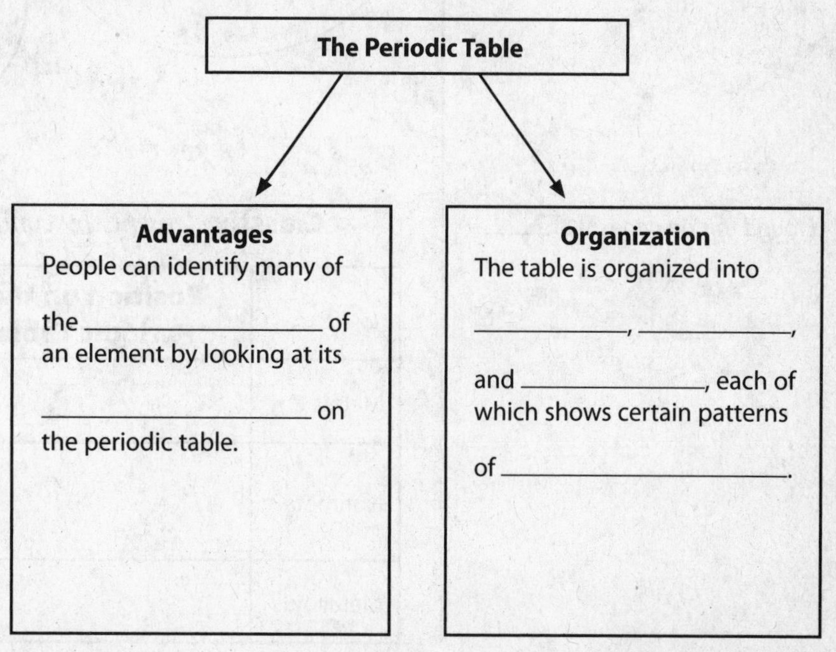

Main Idea	Details

I found this on page ______.

Interpret *the symbols on the element key below. Identify what each symbol stands for.*

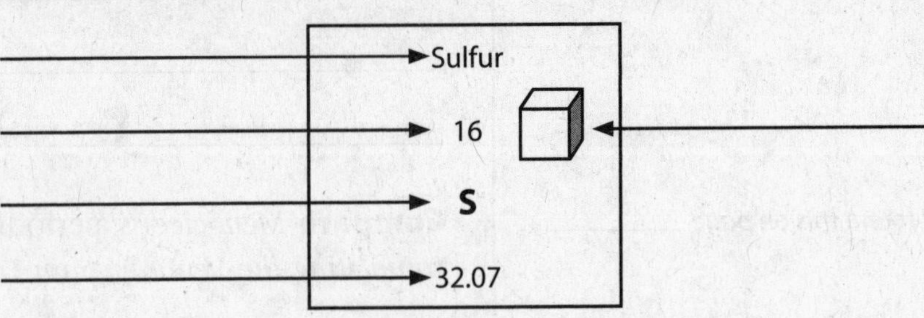

I found this on page ______.

Organize *information about how the* periodic table *is arranged by completing the concept map.*

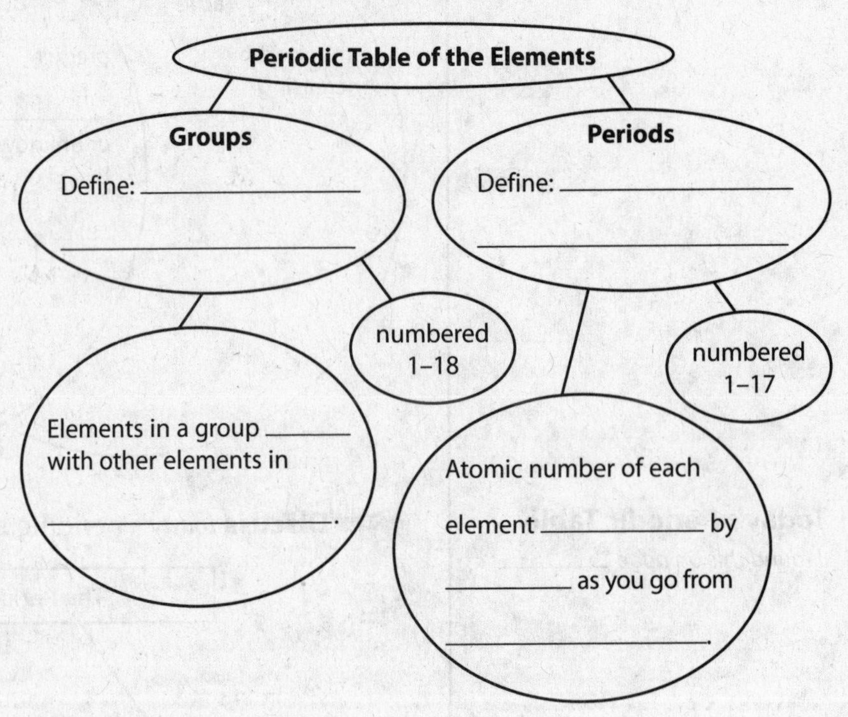

I found this on page ______.

Classify *elements by completing the chart.*

	Position on the Periodic Table	Properties
Metals		
Nonmetals		
Metalloids		

Main Idea | **Details**

I found this on page ________.

Identify *the regions of the* periodic table *as metals, nonmetals, or metalloids. Color each region with a different color. Label each area that you shade.*

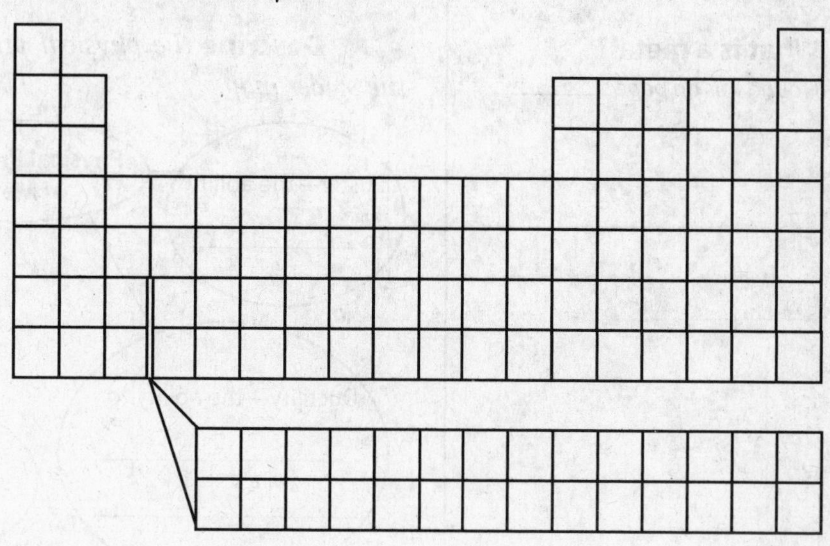

How Scientists Use the Periodic Table

I found this on page ________.

Assess *information about the* periodic table. *Read the statement below. If the statement is true, write* true *on the line. If it is false, rewrite the underlined portion of the statement so that it is true.*

When scientists produce a few atoms of a synthetic element in the laboratory, <u>they have no way to determine the element's properties.</u>

____

____

Analyze It Earth's atmosphere is composed mostly of nitrogen and oxygen. In order to float, a balloon has to be filled with something that is lighter than air. Use the information on the periodic table to find out which elements might be suitable for making a balloon float. Explain your choice or choices.

____

____

____

____

____

____

____

Lesson 2 Metals

Scan *Lesson 2 in your book. In your Science Journal, write three questions you have about metals. Try to answer your questions as you read.*

Main Idea	Details
What is a metal? *I found this on page* ______.	**Describe** *the physical properties of* metals *by completing the spider map.*

Physical Properties of Metals
- Luster—the ability to ______
- Conductivity—the ability to conduct ______ and ______
- Ductility—the ability to ______ ______
- Malleability—the ability to ______ ______

I found this on page ______.

Identify *other physical properties of* metals.

1. ______
2. ______
3. ______ ______

Group 1: Alkali Metals
I found this on page ______.

Name *the 6* alkali metals.

Alkali metals are the elements in group ____ of the periodic table. The six alkali metals are ______, ______, ______, ______, ______, and ______.

I found this on page ______.

Assess *information about* alkali metals. *Circle the correct choice in each set of parentheses.*

Characteristics of Alkali Metals

- react (quickly, slowly) with other elements
- found in nature (as elements, in compounds)
- have a (dull, shiny) appearance
- (soft, hard)
- have the (highest, lowest) densities of all metals

Main Idea	Details

Group 2: Alkaline Earth Metals

I found this on page ________.

Compare and contrast *the* alkali metals *and the* alkaline earth metals *by completing the Venn diagram.*

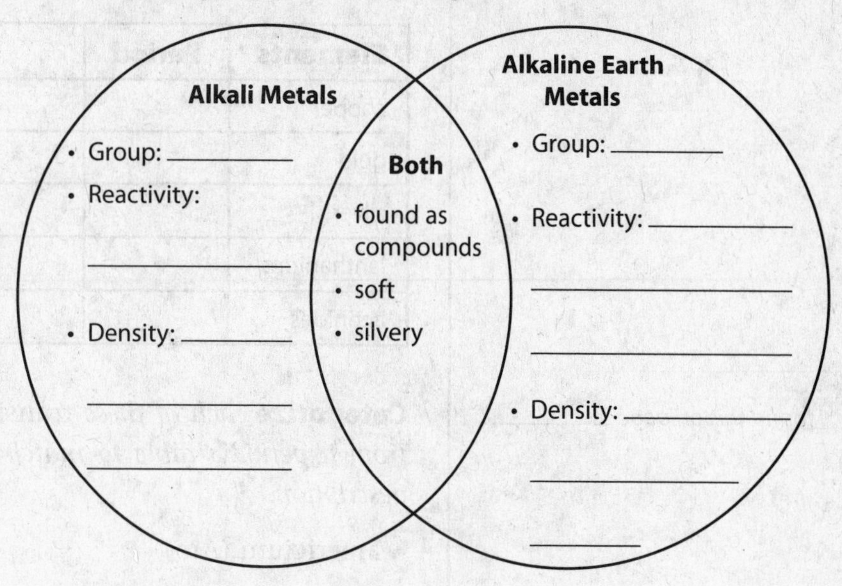

Groups 3–12: Transition Elements

I found this on page ________.

Organize information *about* transition elements *by completing the table.*

Location on the periodic table	Properties	Uses

I found this on page ________.

Describe *the lanthanide and actinide series.*

____

____

____

____

Main Idea | Details

I found this on page ______.

Identify *the periods and uses of the* transition elements *in the chart.*

Elements	Period	Uses
copper		
gold		
silver		
lanthanides		
actinides		

I found this on page ______.

Categorize *each of these* transition elements. *Use information from a periodic table to match each element's symbol with its description.*

- americium (Am)
- lead (Pb)
- francium (Fr)
- mercury (Hg)

Symbol	Description
	Least metallic of the four elements listed
	Metal that has 95 protons
	Element that is more metallic than cesium
	Metal that is liquid at room temperature

Patterns in Properties of Metals

I found this on page ______.

Describe *the arrangement of* metals *in the periodic table in terms of their properties.*

______

______

Analyze It Which element are you more likely to find as a free element rather than a compound, lead or calcium? Explain how using the periodic table can help answer this question.

______

______

______

______

______

Lesson 3 Nonmetals and Metalloids

Predict *three facts you will learn about in Lesson 3. Look at the illustrations in the lesson to give you some clues. Write your facts in your Science Journal.*

Main Idea	Details

The Elements of Life

I found this on page ________.

Discuss *the elements of life. Cross out the incorrect words in the parentheses.*

The human body is made up of mostly (metals, nonmetals).

Ninety-six percent of the mass of the human body is composed of (oxygen, nitrogen, iron, carbon, silicon, copper), and hydrogen.

How are nonmetals different from metals?

I found this on page ________.

Explain nonmetals *by completing the spider map.*

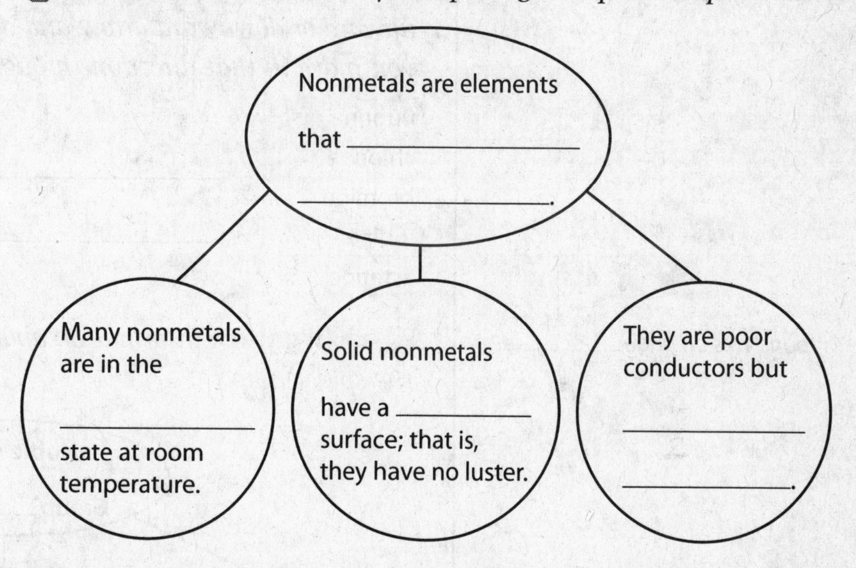

I found this on page ________.

Indicate *which elements in groups 14–16 are* nonmetals. *Classify each of these elements as a solid or gas at room temperature.*

Group 14	Group 15	Group 16

I found this on page ________.

Recall *the feature of the periodic table that helps to locate* nonmetals.

____

____

Main Idea	Details
I found this on page ________.	**Describe** *the* halogens *by completing the graphic organizer.*

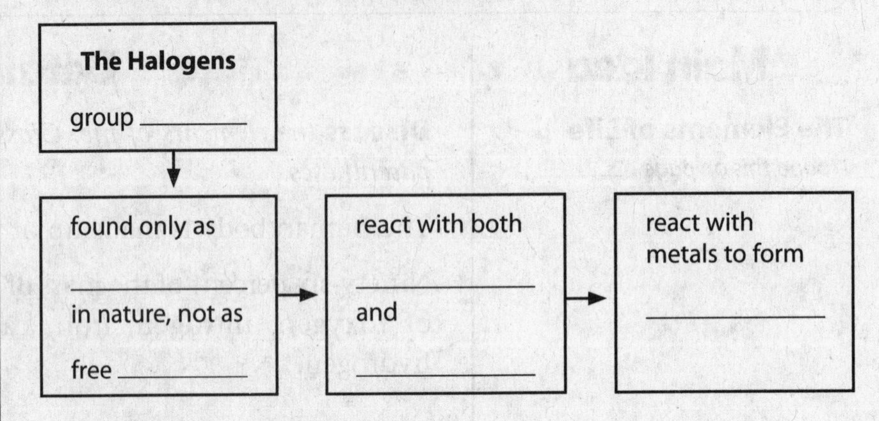

I found this on page ________.

Analyze *the properties of the elements in group 17. Draw an upward or downward arrow and describe how reactivity changes as you move in that direction through the group.*

fluorine
chlorine ________________________________
bromine
iodine ________________________________
astatine

I found this on page ________.

Organize *information about the* noble gases *by completing the spider map.*

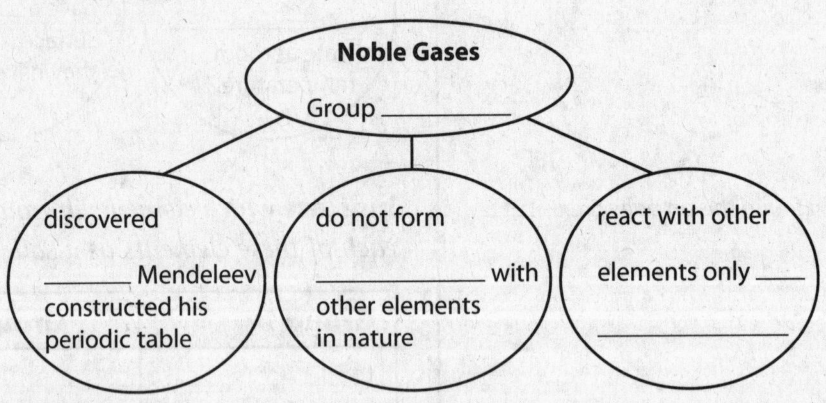

I found this on page ________.

Explain *how hydrogen shows properties of both metals and* nonmetals.

____

____

____

____

____

Main Idea	Details

Metalloids

I found this on page ______.

Define metalloid.

______

______

Classify *characteristics of* metalloids *as like metals and like* nonmetals.

Metalloids	
Like Metals	**Like Nonmetals**
conduct electricity at ______ temperatures	stop electricity from flowing at ______ temperatures

I found this on page ______.

Define semiconductor, *and tell how it is useful.*

______

______

______

______

______

Metals, Nonmetals, and Metalloids

I found this on page ______.

Explain *how knowing the position of an element on the periodic table can help you find a proper use for an element.*

______

______

______

______

______

______

Connect It Without actually seeing the elements themselves, what can you infer from the positions of polonium and bismuth on the periodic table? How reactive do you think they might be? What do you think they might look like?

______

______

______

Review The Periodic Table

Chapter Wrap-Up

Now that you have read the chapter, think about what you have learned. Complete the **What I Learned** *column of the K-W-L chart at the beginning of this chapter.*

Use this checklist to help you study.

- ❑ Complete your Foldables® Chapter Project.
- ❑ Study your *Science Notebook* on this chapter.
- ❑ Study the definitions of vocabulary words.
- ❑ Reread the chapter, and review the charts, graphs, and illustrations.
- ❑ Review the Understanding Key Concepts at the end of each lesson.
- ❑ Look over the Chapter Review at the end of the chapter.

THE BIG IDEA

Summarize It Reread the chapter Big Idea and the lesson Key Concepts. Think of the periodic table as a map, with the top being north, the bottom south, the right side east, and the left side west. How would you describe the locations of hydrogen, the alkali metals, metals, nonmetals, and metalloids?

Challenge *Create your own periodic table. Organize something other than the elements. Choose a group of items that might exhibit repeating, predictable patterns of characteristics. List those characteristics, and sort the items into columns and rows. Some possible items for your periodic table might be music or food.*

Name ______________________ Date __________

Using Energy and Heat

THE BIG IDEA

What are energy transfers and energy transformations?

Before You Read

Before you read the chapter, think about what you know about energy and heat. Record three things that you already know about energy transfers and transformations in the first column. Then write three things that you would like to learn about this topic in the second column. Complete the final column of the chart when you have finished the chapter.

K What I Know	W What I Want to Learn	L What I Learned

Chapter Vocabulary

Lesson 1	Lesson 2	Lesson 3
NEW energy potential energy chemical energy nuclear energy kinetic energy electric energy mechanical energy thermal energy wave sound energy radiant energy **REVIEW** speed	**NEW** law of conservation of energy energy transfer energy transformation work open system closed system renewable energy resource nonrenewable energy resource **ACADEMIC** resource	**NEW** temperature heat conduction radiation convection vaporization thermal conductor thermal insulator

Lesson 1 Forms of Energy

Scan *Lesson 1. Read the lesson titles and bold words. Look at the pictures. Identify three facts you discovered about forms of energy. Record your facts in your Science Journal.*

Main Idea	Details
Energy *I found this on page ________.*	**Define** energy, *and provide examples of its influences.*

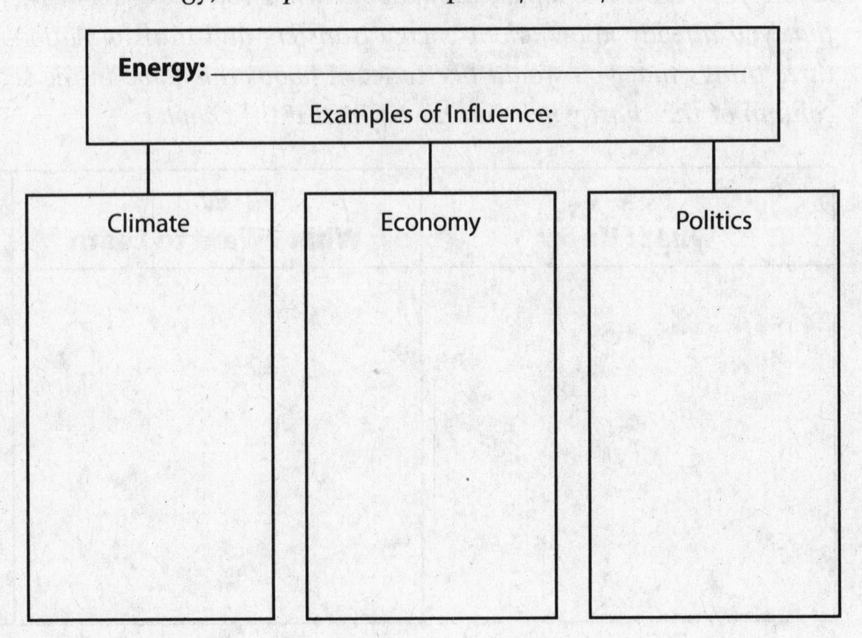

Main Idea	Details
Potential Energy *I found this on page ________.*	**Describe** potential energy.
I found this on page ________.	

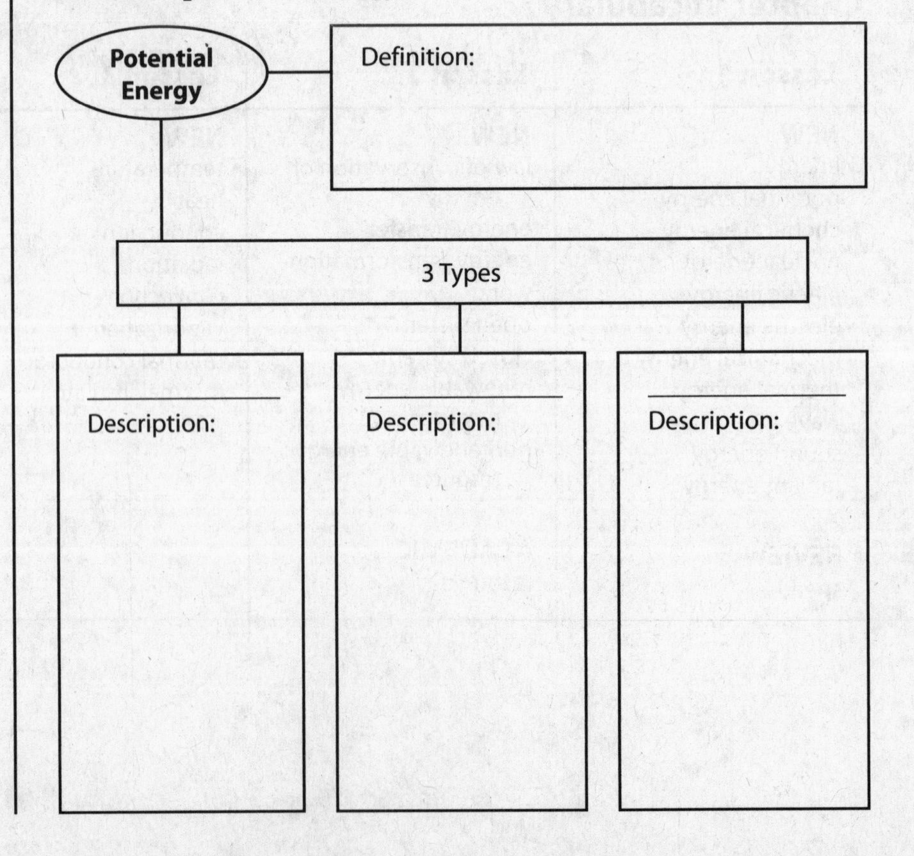

Main Idea	Details
Kinetic Energy *I found this on page* ______.	**Differentiate** potential energy *and* kinetic energy. Potential energy: ______ ______ Kinetic energy: ______ ______
I found this on page ______.	**Relate** kinetic energy *to mass and speed of objects.* No speed → ______ kinetic energy Greater mass → ______ kinetic energy Greater speed → ______ kinetic energy
I found this on page ______.	**Explain** *why* electric energy *is a form of* kinetic energy. ______ ______ ______
Combined Kinetic Energy and Potential Energy *I found this on page* ______.	**Model** *a system and an environment. Define each term, and label the part of the diagram that represents each term.* System: ______ ______ Environment: ______ ______

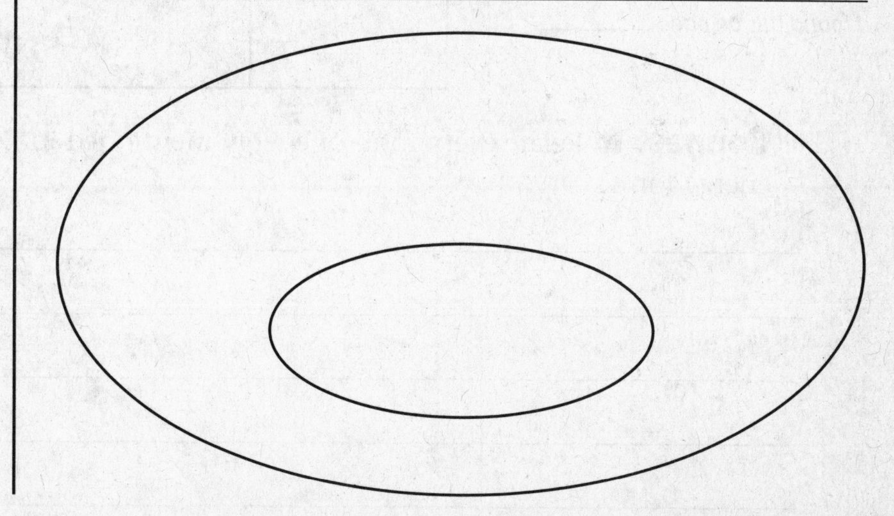

Main Idea

Details

I found this on page ________.

Compare and contrast mechanical energy *and* thermal energy.

Mechanical Energy	Both	Thermal Energy

Energy Carried by Waves

I found this on page ________.

Characterize waves.

Wave
A disturbance

does move ________
from one place to another.

does not move
________.

moves in which direction?
________.

Identify *two types of energy carried by* waves.

I found this on page ________.

I found this on page ________.

Type of Energy	Carried By	Moves Through

Connect It Relate every form of energy mentioned in Lesson 1 to at least one other energy form.

____

____

____

____

____

Lesson 2 Energy Transfers and Transformations

Predict *three facts that will be discussed in Lesson 2 after reading the headings. Write your facts in your Science Journal.*

Main Idea	Details
Law of Conservation of Energy *I found this on page* ______.	**Express** *the* law of conservation of energy *in a diagram.*

Energy
- can be
- cannot be

I found this on page ______.

Apply *the* law of conservation of energy *to* energy transformations *that occur in a flashlight.*

Chemical energy from the battery → transformed to → ______ moving through the contact strip to the bulb → transformed to → ______ + ______ in the bulb

I found this on page ______.

Contrast energy transfer *with* energy transformation.

Energy Transfer	Energy Transformation

I found this on page ______.

Identify *three criteria required for an* energy transfer *to be considered an occurrence of* work.

1. ______
2. ______
3. ______

I found this on page ______.

Explain *what makes an* energy transformation *inefficient.*

______

Main Idea | Details

I found this on page ________.

Distinguish *an* open system *from a* closed system.

Open System	Closed System
Definition: Example: Explanation:	Definition: Example: Explanation:

Energy Transformations and Electric Energy

I found this on page ________.

Categorize *resources that people transform into electric energy.*

I found this on page ________.

Category	Definition	Types
	an energy resource that is replaced as fast as, or faster than, it is used	• • • • •
Nonrenewable energy resource		• •

I found this on page ________.

Designate *what percentages of our electric energy come from categories of resources. Write the percentage in the category boxes in the table above.*

Synthesize It Explain two reasons it is wise to use energy-efficient electric appliances.

____

____

____

____

Lesson 3 Particles in Motion

Skim *Lesson 3 in your book. Read the headings and look at the photos and illustrations. Identify three things you want to learn more about as you read the lesson. Record your ideas in your Science Journal.*

Main Idea	Details
Kinetic Theory *I found this on page* ________.	**Record** *3 major points of the kinetic theory.* **1.** ________ **2.** ________ **3.** ________ ________
I found this on page ________.	**Relate** temperature *to thermal expansion.* When temperature increases, the ________ ________ of particles ________. Thermal expansion occurs when particles ________ ________ and ________.
I found this on page ________.	**Differentiate** temperature *and* heat. *Circle the word in each definition that most distinguishes the meaning of the term. (Hint: They both start with the letter "M.")* Temperature: ________ ________
I found this on page ________.	Heat: ________ ________
I found this on page ________.	**Identify and describe** *the condition under which two materials in contact would each have* temperature *but no* heat. ________ ________ ________

Main Idea	Details

Heat Transfer

I found this on page ________.

Describe *3 ways in which thermal energy is transferred.*

Method	Description
Conduction	

Heat and Changes of State

I found this on page ________.

I found this on page ________.

Characterize *the relationship between* heat *and typical changes of state.*

Add enough thermal energy to a solid…	
Remove enough thermal energy from a liquid…	
	Vaporization, or the change from liquid to gas, occurs.
	Condensation, or the change from gas to liquid, occurs.

Main Idea	Details
I found this on page ______.	**Differentiate** *types of* vaporization.

Two Types of Vaporization

______ occurs ______	______ occurs ______

I found this on page ______.

Identify *which of the processes above requires an addition of thermal energy.*

______

I found this on page ______.

Distinguish *sublimation from deposition. Circle the process that represents an increase in thermal energy.*

Sublimation	Deposition

Conductors and Insulators

I found this on page ______.

Contrast *the ability of materials to transfer thermal energy.*

Thermal Conductor → Energy transfer:
Reason:

Thermal Insulator → Energy transfer:
Reason:

Connect It Describe the thermal energy transfers between particles that occur when you overheat some soup for lunch, then drop an ice cube in it to cool it off.

______

______

______

______

Review

Using Energy and Heat

Chapter Wrap-Up

Now that you have read the chapter, think about what you have learned. Complete the final column in the chart on the first page of this chapter.

Use this checklist to help you study.

- ❑ Complete your Foldables® Chapter Project.
- ❑ Study your *Science Notebook* on this chapter.
- ❑ Study the definitions of vocabulary words.
- ❑ Reread the chapter, and review the charts, graphs, and illustrations.
- ❑ Review the Understanding Key Concepts at the end of each lesson.
- ❑ Look over the Chapter Review at the end of the chapter.

THE BIG IDEA

Summarize It Reread the chapter Big Idea and the lesson Key Concepts. Summarize what you have learned by converting two of the Key Concept questions from each lesson into factual answers.

Lesson 1 (three Key Concepts): ______________________________

____

____

____

Lesson 2 (three Key Concepts): ______________________________

____

____

____

Lesson 3 (three Key Concepts): ______________________________

____

____

____

____

Challenge *Think of yourself riding a bike around your neighborhood as a system. Describe all of the energy transfers and transformations that occur in that system. Write your description in a summary to share in your class.*

Name ____ Date ______________

The Earth System

How do Earth systems recycle Earth materials?

Before You Read

Before you read the chapter, think about what you know about Earth systems. Record three things that you already know about Earth systems in the first column. Then write three things that you would like to learn about in the second column. Complete the final column of the chart when you have finished the chapter.

K What I Know	W What I Want to Learn	L What I Learned

Chapter Vocabulary

Lesson 1	Lesson 2
NEW carbon cycle greenhouse gas phosphorus cycle **REVIEW** fossil fuels	**NEW** luster streak cleavage fracture crust mantle lithosphere asthenosphere core **ACADEMIC** structure

Lesson 1 Earth Systems and Interactions

Scan *Lesson 1. Read the lesson titles and bold words. Look at the pictures. Identify three facts you discovered about Earth systems. Record your facts in your Science Journal.*

Main Idea | Details

Earth Systems

I found this on page ______.

Describe *4 smaller systems that make up the larger Earth system.*

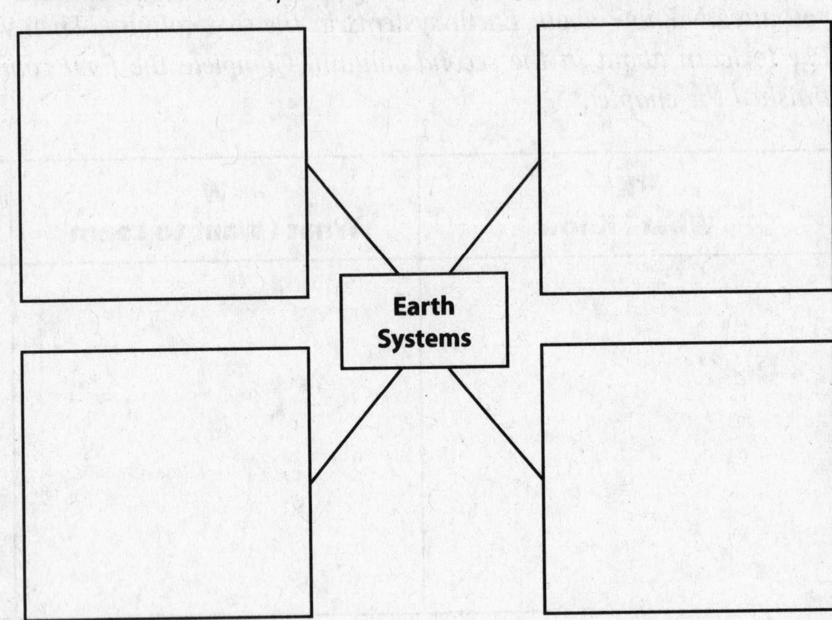

I found this on page ______.

Interactions Among Earth Systems

I found this on page ______.

Explain *how Earth systems interact in two cycles.*

Cycle	Interactions
Water cycle	
Rock cycle	

I found this on page ______.

Generalize *what happens to the amount of material cycling through each of Earth's systems.*

______

______

______

Main Idea	Details
I found this on page ______.	**Demonstrate** *movement of carbon in the* carbon cycle. *Provide examples of carbon moving between Earth systems.*

From the geosphere	carbon ⇒	To the atmosphere
Example:		
From the atmosphere	carbon ⇒	To the biosphere
Example:		
From the atmosphere	carbon ⇒	To the hydrosphere
Example:		
From the biosphere	carbon ⇒	To the atmosphere
Example:		
From the hydrosphere	carbon ⇒	To the geosphere
Example:		

I found this on page ______.

Generalize *what happens to the total amount of carbon as it cycles through Earth's systems.*

____

____

Main Idea	Details

I found this on page ______.

Order *carbon reservoirs by the proportions of Earth's carbon that they contain.*

Most carbon
Least carbon

I found this on page ______.

Identify *two activities that change the amount of CO_2 in the atmosphere over time.*

Natural Activity	Human Activity

I found this on page ______.

Relate greenhouse gases *to global warming.*

Levels of atmospheric CO_2 increase. → More ______ ______ is ______ ______. → Earth's average temperature ______.

I found this on page ______.

Characterize *phosphorus.*

Phosphorus

Earth system NOT found in:

Earth systems found in:
-
-
-

Not found as an element, but as compounds called:

Main Idea	Details

I found this on page ________.

Demonstrate *movement of phosphorus in the* phosphorus cycle. *Provide examples of phosphorus moving between Earth systems.*

From the geosphere	phosphorus →	To the hydrosphere
Example:		
From the hydrosphere	phosphorus →	To the geosphere
Example:		
From the geosphere	phosphorus →	To the biosphere
Example:		

I found this on page ________.

Contrast *the rate at which phosphorus cycles through different Earth systems.*

Geosphere and Hydrosphere	Biosphere

I found this on page ________.

Express *how two human activities disturb the* phosphorus cycle.

1. ____

____

2. ____

____

Synthesize It You are part of the biosphere. How will you interact with the atmosphere, the hydrosphere, and the geosphere today?

____

____

____

____

Lesson 2 # The Geosphere

Predict *three facts that will be discussed in Lesson 2 after reading the headings. Write your facts in your Science Journal.*

Main Idea	Details
Materials in the Geosphere *I found this on page ______.*	**Order** *the layers of Earth material.*

Surface
Core

I found this on page ______.

Characterize *minerals.*

Minerals

I found this on page ______.

Describe *properties used to identify minerals.*

Property	Description
Luster	
Streak	
Cleavage	
Fracture	

Main Idea	Details
I found this on page ______.	**Explain** *why calcite is more likely to interact with other minerals than is quartz.*
I found this on page ______.	**Compare and contrast** *rocks and minerals.*

Mineral

Both

naturally occurring solids

Rock

Main Idea	Details
I found this on page ______.	**Distinguish** *3 main types of rocks.*

Type	Description
Igneous	
Metamorphic	
Sedimentary	

Main Idea	Details
I found this on page ______.	**Infer** *why it is important to protect soil.*

Main Idea | Details

I found this on page ________.

Assess *how interactions among all Earth systems occur in soil.*

____

____

I found this on page ________.

Differentiate *layers of soil.*

	A-horizon	B-horizon	C-horizon
Description			

Structure of the Geosphere

I found this on page ________.

Describe *the structure of the geosphere below Earth's* crust.

Layer	Description	Density
Upper mantle		
Lower mantle		
Outer core		
Inner core		

I found this on page ________.

Relate *the density of material in Earth's layers to the structure of the geosphere.*

____
pulled
____

through
____
toward
____________________.

Main Idea	Details

I found this on page ______.

Characterize *Earth's* crust.

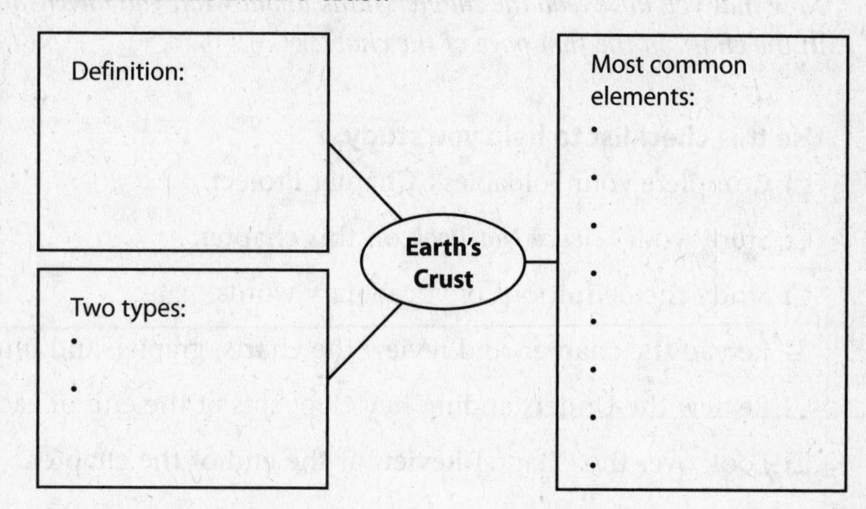

I found this on page ______.

Define *Earth's* mantle.

____

I found this on page ______.

Contrast *the* lithosphere *and the* asthenosphere.

Lithosphere	Asthenosphere

I found this on page ______.

Identify *differences in the layers of Earth's* core.

Layer	Outer Core	Inner Core
State (circle and explain)	solid or liquid Why?	solid or liquid Why?

Connect It Describe the main difference between the core and the other layers of the geosphere. Then describe a characteristic of soil that distinguishes it from other materials in the geosphere.

____

____

____

Review **The Earth System**

Chapter Wrap-Up

Now that you have read the chapter, think about what you have learned. Complete the final column in the chart on the first page of the chapter.

Use this checklist to help you study.

- ❑ Complete your Foldables® Chapter Project.
- ❑ Study your *Science Notebook* on this chapter.
- ❑ Study the definitions of vocabulary words.
- ❑ Reread the chapter, and review the charts, graphs, and illustrations.
- ❑ Review the Understanding Key Concepts at the end of each lesson.
- ❑ Look over the Chapter Review at the end of the chapter.

THE BIG IDEA **Summarize It** Reread the chapter Big Idea and the lesson Key Concepts. Summarize how Earth's systems drive changes on and beneath Earth's surface.

Challenge *Make a three-dimensional model cross-section of Earth. Include not only the layers of the geosphere, but also represent the hydrosphere, the atmosphere, and the biosphere. Share your model with your class.*

Name ______________________________ Date ____________

Earth's Changing Surface

How do natural processes change Earth's surface over time?

Before You Read

Before you read the chapter, think about what you know about Earth's surface and how it changes. Record your ideas in the first column. Pair with a partner, and discuss his or her thoughts. Write those ideas in the second column. Then record what you both would like to share with the class in the third column.

Think	Pair	Share

Chapter Vocabulary

Lesson 1	Lesson 2	Lesson 3
NEW	**NEW**	**NEW**
plate tectonics	earthquake	weathering
continental drift	fault	erosion
convergent boundary	mid-ocean ridge	physical weathering
divergent boundary	hot spot	chemical weathering
transform boundary	lava flow	soil
subduction zone	volcanic ash	sediment
compression	caldera	deposition
tension		
shear		**REVIEW**
		composition

Lesson 1 Plate Tectonics

Skim *Lesson 1 in your book. Read the headings and look at the photos and illustrations. Identify three things you want to learn more about as you read the lesson. Record your ideas in your Science Journal.*

Main Idea	Details

Plate Motion

I found this on page ______.

Differentiate plate tectonics *from* continental drift.

Plate Tectonics	Continental Drift

I found this on page ______.

Explain *evidence of* plate tectonics.

Evidence	Explanation
Shape of continents	
Fossil evidence	
Geological evidence	

I found this on page ______.

Relate *convection to the movement of continents. Include the words in parentheses in your explanations.*

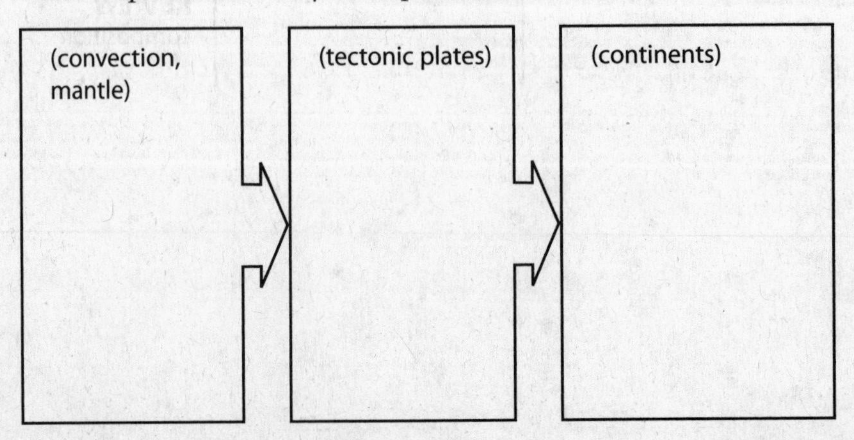

Main Idea	Details

Tectonic Plate Boundaries

Contrast *three main types of plate boundaries.*

I found this on page ________.

I found this on page ________.

Type	Description
Convergent	
Divergent	
Transform	

I found this on page ________.

Define subduction zone, *and circle the boundary in the table above where* subduction zones *occur.*

Forces Changing Earth's Surface

I found this on page ________.

Diagram *forces present at plate boundaries, and write examples of how they change Earth's surface.*

	Compression	Tension	Shear
Diagram			
Change			a shifted road

Connect It Suppose you could time-travel 50 million years into the future. If you took a land map with you, would it be useful? Explain.

Lesson 2 Earthquakes and Volcanoes

Scan *Lesson 2. Read the lesson titles and bold words. Look at the pictures. Identify three facts you discovered about earthquakes and volcanoes. Record your facts in your Science Journal.*

Main Idea	Details

Earthquakes

I found this on page ______.

Sequence *the events that cause* earthquakes.

1. Forces push on rock layers along a fault.
2.
3.
4.

I found this on page ______.

Generalize *the locations at which most* earthquakes *occur.*

____

I found this on page ______.

Record *changes caused by* earthquakes. *After you complete the organizer, circle the factor that is responsible for the most damage to buildings.*

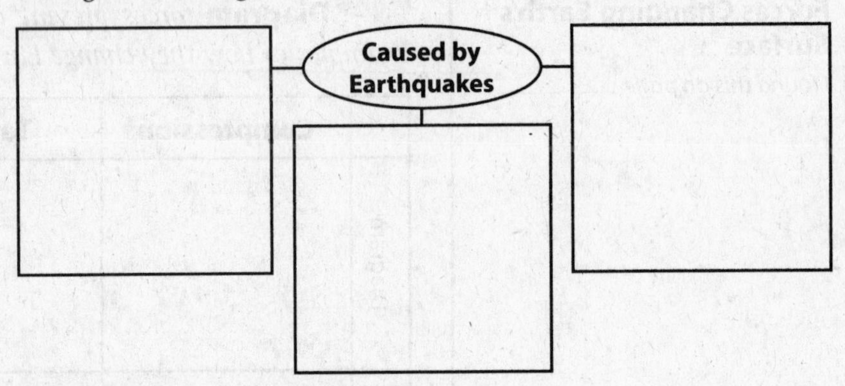

Volcanoes

I found this on page ______.

Identify *events that cause volcanoes to form. Circle the location along which* mid-ocean ridges *form.*

Location	Event
Divergent boundary	
Convergent boundary	
Hot spots	

Main Idea	Details
I found this on page ______.	**Assess** *the reason magma forms as rocks move from deep inside Earth toward Earth's surface.*
I found this on page ______.	**Contrast** *changes caused by volcanoes.*

Event	Description
Lava flow	
Explosive eruption	

I found this on page ______.

Diagram *and label the shapes of two types of volcanoes.*

I found this on page ______.

Sequence *the formation of a* caldera.

Magma builds up in a magma chamber. → ______ → ______

I found this on page ______.

Describe *effects of volcanoes on the atmosphere.*

Analyze It Explain why earthquakes and volcanoes often occur in the same places.

Lesson 3 Weathering, Erosion, and Deposition

Predict *three facts that will be discussed in Lesson 3 after reading the headings. Record your predictions in your Science Journal.*

Main Idea / Details

Weathering

I found this on page ______.

Describe *how* weathering *and* erosion *work together to wear down mountains.*

____

____

____

I found this on page ______.

I found this on page ______.

Distinguish *types of* weathering.

Type	Description
Physical weathering	
Chemical weathering	
Interaction between the two	

I found this on page ______.

Relate weathering *and biological activity to the formation of* soil.

Weathering → Soil

Biological activity → Soil

Main Idea	Details

Erosion

I found this on page ________.

Differentiate *processes that change Earth's surface. Include the term* sediment *in your explanation.*

Process	Description
Erosion	
by water	
by ice	
by wind	
Deposition	
by water	
by ice	
by wind	

Deposition

I found this on page ________.

I found this on page ________.

Generalize *the movement of* sediment *in the* erosion-deposition *cycle.*

____

____

Connect It Describe how organisms play a role in the cycle of weathering, erosion, and deposition.

____

____

____

____

Review

Earth's Changing Surface

Chapter Wrap-Up

Now that you have read the chapter, think about what you have learned.

Use this checklist to help you study.

- ❑ Complete your Foldables® Chapter Project.
- ❑ Study your *Science Notebook* on this chapter.
- ❑ Study the definitions of vocabulary words.
- ❑ Reread the chapter, and review the charts, graphs, and illustrations.
- ❑ Review the Understanding Key Concepts at the end of each lesson.
- ❑ Look over the Chapter Review at the end of the chapter.

THE BIG IDEA

Summarize It Reread the chapter Big Idea and the lesson Key Concepts. Referencing Lessons 1, 2, and 3, describe the many things that could happen to rock on a mountain top over a very long period of time.

Challenge *Research news stories to learn about a recent natural event that significantly changed Earth's surface. Design a poster about the event that includes both pictures and written descriptions. Present the poster to your class.*

Name ____________________ Date __________

Using Natural Resources

How can people protect Earth's resources?

Before You Read

Before you read the chapter, think about what you know about Earth's resources. Record three things that you already know about Earth's resources in the first column. Then write three things that you would like to learn about how these resources can be protected in the second column. Complete the final column of the chart when you have finished the chapter.

K What I Know	W What I Want to Learn	L What I Learned

Chapter Vocabulary

Lesson 1	Lesson 2	Lesson 3
NEW natural resource nonrenewable resource renewable resource inexhaustible resource geothermal energy **REVIEW** atmosphere	**NEW** pollution ozone layer photochemical smog global warming acid precipitation **ACADEMIC** occur	**NEW** sustainability recycling

Lesson 1 Earth's Resources

Scan *Lesson 1. Read the lesson titles and bold words. Look at the pictures. Identify three facts you discovered about Earth's resources. Record your facts in your Science Journal.*

Main Idea | Details

Natural Resources

I found this on page ________.

Characterize natural resources.

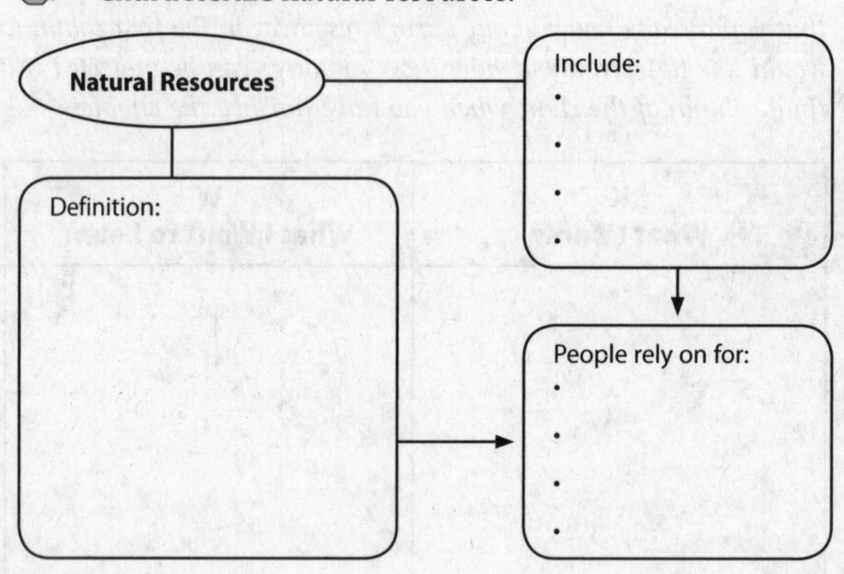

Nonrenewable Resources

I found this on page ________.

I found this on page ________.

Distinguish *two types of* nonrenewable resources. *Record three examples and uses of each.*

	Examples	Uses
Fossil Fuels		
Minerals		

Main Idea	Details

Renewable Resources

I found this on page ______.

I found this on page ______.

Examine *uses of three types of* renewable resources.

	Examples	Use
Air	oxygen	
	carbon dioxide	
Land	fertile soil	
	wildlife ecosystems	
Water	freshwater	
	oceans	

Inexhaustible Resources

I found this on page ______.

I found this on page ______.

I found this on page ______.

Identify *three* inexhaustible resources *that can be used by power plants to provide electricity used in homes.*

1. ______
2. ______
3. ______

Contrast *three types of* natural resources.

Nonrenewable	Renewable	Inexhaustible

Analyze It Identify an example of a nonrenewable resource, a renewable resource, and an inexhaustible resource in the room you are in right now.

______

______

______

______

Lesson 2 Pollution

Predict *three facts that will be discussed in Lesson 2 after reading the headings. Record your predictions in your Science Journal.*

Main Idea / Details

What is pollution?

I found this on page ________.

Relate *causes to effects in examples of* pollution.

Cause	Effect
Air Source:	Effect:
Land Source:	Effect:
Water Source:	Effect:

Air Pollution

I found this on page ________.

I found this on page ________.

Organize *factors related to air* pollution.

Air Pollution Factors

Ozone Loss	**Photochemical Smog**
Cause:	Cause:
Effect:	Effect:

Global Warming	**Acid Precipitation**
Cause:	Cause:
Effect:	Effect:

Main Idea	Details

Water Pollution
I found this on page ________.

Identify *examples of sources of water* pollution.

Sources of Water Pollution

I found this on page ________.

Assess *two harmful consequences of water* pollution.

1. ________________
2. ________________

Land Pollution
I found this on page ________.

Describe *examples of land* pollution.

Source	Pollution and Its Effect
Homes	
Farms	
Industry	
Mining	

Synthesize It Describe how one source of pollution mentioned in Lesson 2 can affect all three parts of the environment—air, water, and land.

____

____

____

____

Lesson 3 Protecting Earth

Skim *Lesson 3 in your book. Read the headings and look at the photos and illustrations. Identify three things you want to learn more about as you read the lesson. Record your ideas in your Science Journal.*

Main Idea	Details

Monitoring Human Impact on Earth

I found this on page ________.

Organize *information about ways that scientists collect data on environmental conditions.*

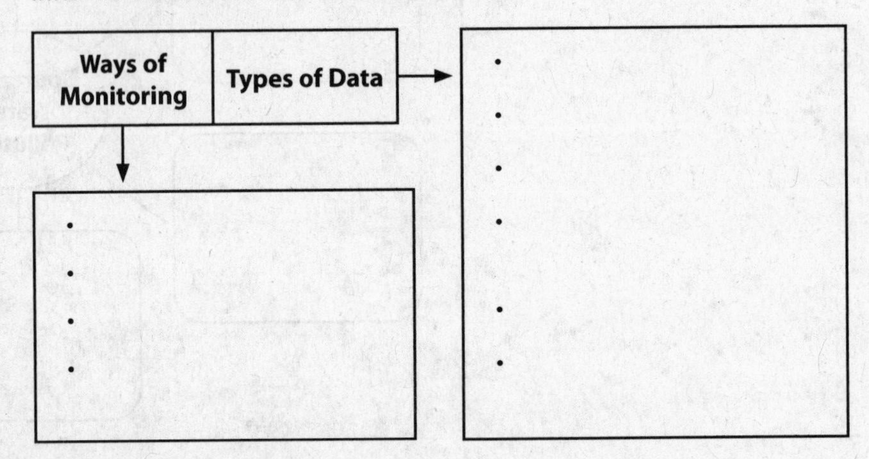

Developing Technologies

I found this on page ________.

Assess *the benefits of technologies for saving resources with reduced pollution.*

	Save Resources	Reduce Pollution
Water	low-flow shower heads and toilets	
Energy	compact fluorescent lightbulbs	

I found this on page ________.

Note *benefits of using CFC replacements and alternative fuels.*

CFC replacements: ________________________

Alternative fuels: ________________________

____

Main Idea	Details

I found this on page ______.

Differentiate *automobile technologies.*

HEV	FCV

Making a Difference
I found this on page ______.

Apply *the concept of* sustainability.

Meet present needs in ways that	

I found this on page ______.

Describe *examples of individual efforts that contribute to environmental* sustainability.

Type of Action	Example Description
Restoring	
Rethinking	
Reducing	
Reusing	
Recycling	

Connect It Explain how taking notes on the back of a worksheet in class contributes to sustainability. What else can you do with the paper to promote sustainability?

____

____

____

____

Review Using Natural Resources

Chapter Wrap-Up

Now that you have read the chapter, think about what you have learned. Complete the final column in the chart on the first page of the chapter.

Use this checklist to help you study.

- ❑ Complete your Foldables® Chapter Project.
- ❑ Study your *Science Notebook* on this chapter.
- ❑ Study the definitions of vocabulary words.
- ❑ Reread the chapter, and review the charts, graphs, and illustrations.
- ❑ Review the Understanding Key Concepts at the end of each lesson.
- ❑ Look over the Chapter Review at the end of the chapter.

THE BIG IDEA **Summarize It** Reread the chapter Big Idea and the lesson Key Concepts. Describe how the world would be changed if all the power plants that produce electricity from fossil fuels could be replaced with power plants that generate electricity from solar energy and wind power.

____

____

____

____

____

____

____

____

____

____

____

____

Challenge *Do research about products using new technologies that are being developed to conserve energy and reduce pollution. Make a one-page magazine advertisement that promotes your favorite. Display your ad in your class.*

Name ________________________________ Date ____________

Earth's Atmosphere

How does Earth's atmosphere affect life on Earth?

Before You Read

Before you read the chapter, think about what you know about Earth's atmosphere. Record your thoughts in the first column. Pair with a partner, and discuss his or her thoughts. Write those thoughts in the second column. Then record what you both would like to share with the class in the third column.

Think	Pair	Share

Chapter Vocabulary

Lesson 1	Lesson 2	Lesson 3	Lesson 4
NEW	**NEW**	**NEW**	**NEW**
atmosphere	radiation	wind	air pollution
water vapor	conduction	trade winds	acid precipitation
troposphere	convection	westerlies	photochemical smog
stratosphere	stability	polar easterlies	particulate matter
ozone layer	temperature inversion	jet stream	
ionosphere		sea breeze	
	ACADEMIC	land breeze	
REVIEW	process		
liquid			

Lesson 1 Describing Earth's Atmosphere

Scan *Lesson 1. Read the lesson titles and bold words. Look at the pictures. Identify three facts that you discover about Earth's atmosphere. Record these facts in your Science Journal.*

Main Idea	Details
Importance of Earth's Atmosphere *I found this on page ______.*	**Define** atmosphere, *and identify four things the* atmosphere *does for Earth.* Atmosphere: ______________________ ______________________ **1.** ______________________ ______________________ **2.** ______________________ ______________________ **3.** ______________________ ______________________ **4.** ______________________ ______________________
Origins of Earth's Atmosphere *I found this on page ______.*	**Write** *the number of each event on the time line to describe how Earth's* atmosphere *changed over time.* **1.** Photosynthetic organisms remove carbon dioxide from the air and release oxygen. **2.** Water vapor cools and condenses. Rain falls, evaporates, and eventually accumulates in oceans. **3.** Atmosphere contains present levels of carbon dioxide, oxygen, nitrogen, and other gases. **4.** Atmosphere is mainly water vapor with a little carbon dioxide and nitrogen. **Early atmosphere** → **Present time**

Main Idea	Details

Composition of the Atmosphere

I found this on page ______.

Assess *information about the* atmosphere. *Read each statement below. If the statement is true, write* true *on the line. If the statement is false, write* false *on the line and rewrite the underlined portion so that it is true.*

Earth's atmosphere is mostly made of visible gases, including nitrogen, oxygen, and carbon dioxide.

____

Solid and liquid particles are also present in the atmosphere.

____

I found this on page ______.

Identify *the gases that make up Earth's* atmosphere.

Gases in the Atmosphere	
Percent	**Gas**
78	
21	
1	**a.** **b.** **c.** **d.**

I found this on page ______.

Identify *solid and liquid particles in the* atmosphere.

Particles in the Atmosphere

Solids
- **a.** ______
- **b.** ______
- **c.** ______
- **d.** ______
- **e.** ______

Liquids
- **a.** ______
- **b.** ______ ______
- **c.** ______ ______

Main Idea | **Details**

Layers of the Atmosphere

Describe *the layers of the* atmosphere. *First, list the layers in order from the surface to space. Identify the height of each layer. Then describe each layer.*

Layers of the Atmosphere	
Layer and Height above Earth's Surface	**Description**
________ above 500 km	
Thermosphere	
________ extends from about 50 km to about 85 km	
Stratosphere	
________ from the surface to a height of 8–15 km	

I found this on page ________.

I found this on page ________.

I found this on page ________.

I found this on page ________.

I found this on page ________.

I found this on page ________.

Distinguish ozone *from oxygen.*

Ozone	Oxygen

Main Idea	Details

I found this on page ________.

Identify *the 2 layers of the* atmosphere *that contain the* ionosphere.

1. ________________ **2.** ________________

I found this on page ________.

Explain, *in your own words, how auroras form in the* ionosphere.

____

____

____

____

Air Pressure and Altitude

I found this on page ________.

Describe *the relationship between altitude and air pressure.*

As altitude ______________, air pressure ______________.

Temperature and Altitude

I found this on page ________.

Identify *the changes in temperature and altitude in the different layers of the* atmosphere.

Layer of the Atmosphere	Altitude	Temperature
Troposphere	↑ increases	
Stratosphere	↑ increases	
Mesosphere	↑ increases	
Thermosphere	↑ increases	
Exosphere	↑ increases	

Connect It Suppose that you move from a town near the ocean to a town in the mountains. To what atmospheric changes would your body need to adjust?

____

____

____

____

____

____

Lesson 2 Energy Transfer in the Atmosphere

Predict *three facts that will be discussed in Lesson 2 after reading the headings. Record these facts in your Science Journal.*

Main Idea	Details
Energy from the Sun *I found this on page* ______.	**Define** radiation. Radiation: ______________________ ______________________
I found this on page ______.	**Identify** *the 3 forms of* radiation *that make up most of the Sun's energy.* **1.** ______________________ **2.** ______________________ **3.** ______________________
I found this on page ______.	**Compare and contrast** *infrared and ultraviolet light.*

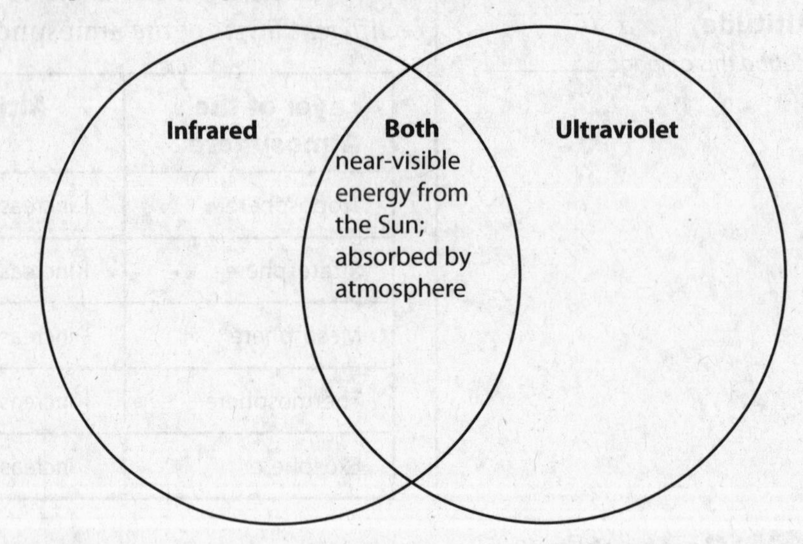

Main Idea	Details
Energy on Earth *I found this on page* ______.	**Color** *the circle graph to represent the portion of* radiation *reflected and absorbed by Earth's surface and atmosphere. Complete the key to show what each color indicates.*

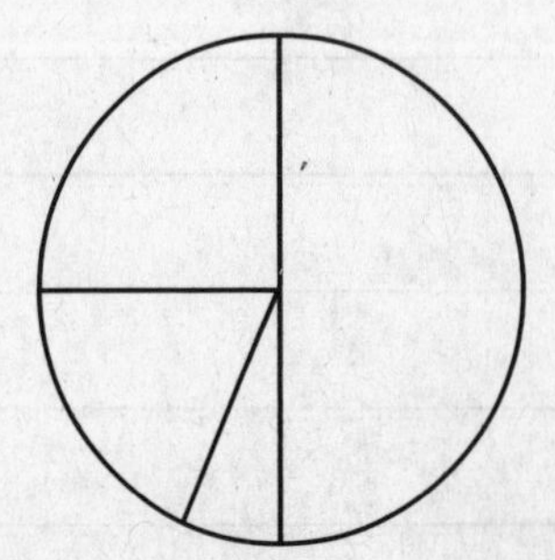

KEY

☐ 25% reflected back to space by particles in the atmosphere
☐ 20% absorbed by particles in the atmosphere
☐ 50% absorbed by Earth's surface
☐ 5% reflected back by land and sea surfaces

Main Idea	Details
Radiation Balance *I found this on page* ________.	**Explain** *how* radiation *levels are kept in balance.* **Incoming** Solar radiation that reaches Earth's surface is ________________ ________________ ________________. **Outgoing** Earth emits ________________ back toward the Sun in the form of ________________.
The Greenhouse Effect *I found this on page* ________.	**Identify** *three greenhouse gases, and include their chemical formulas.* **1.** ________________ **2.** ________________ **3.** ________________
I found this on page ________.	**Draw** *a yellow arrow to indicate the incoming visible light. Draw red arrows to indicate the paths of infrared energy.* 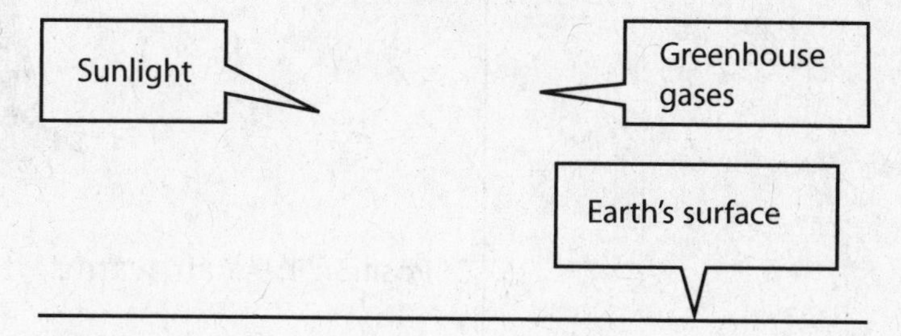
Thermal Energy Transfer *I found this on page* ________.	**Identify** *and define 3 ways that thermal energy is transferred.* **1.** ________________ ________________ **2.** ________________ ________________ **3.** ________________ ________________

Main Idea	Details
I found this on page ______.	**Identify** *each type of energy transfer.*

Sun

surface air

______

______

______

Main Idea	Details
I found this on page ______.	**Describe** *latent heat's relationship to water, and give an example.*

______

______

Example: ______

______

______

Circulating Air

I found this on page ______.

Describe *how air moves as it is heated and cooled. Indicate what happens at each position.*

1

2

3

Position 1: As air warms, it becomes ______ and ______.

Position 2: As air moves away from the warm surface, it loses ______ and ______. Cool air is ______ than warm air, so it begins to ______.

Position 3: Cool air ______ and pushes the ______ air out of the way.

I found this on page ______.

Define stability.

Stability: ______

______

Main Idea | Details

I found this on page ________.

Distinguish *the motion of stable and unstable air.*

Motion of Stable Air	Motion of Unstable Air

I found this on page ________.

Explain *air movement during a thunderstorm.*

During unstable conditions, ground level air is much warmer than ____________. Air rises ____________, cools, and produces large, tall clouds. ____________, released as water vapor, changes from a ____________ to a ____________, adds to the instability, and produces a violent storm.

I found this on page ________.

Sequence *a* temperature inversion.

Ground-level air is nearly the same temperature as ____________________________.

↓

A layer of ____________________ is trapped by a layer of ____________________ above it.

↓

A ____________________ prevents air from mixing and can trap ____________________ in the air close to Earth's surface.

Analyze It While on a picnic in the Rocky Mountains, you notice that clouds form and disappear at the top of the peaks. How can you explain this phenomenon?

____

____

____

____

Lesson 3 Air Currents

Skim *Lesson 3 in your book. Read the headings and look at the photos and illustrations. Identify three things you want to learn more about as you read the lesson. Write your ideas in your Science Journal.*

Main Idea	Details
Global Winds *I found this on page* ________.	**Explain** *the formation of Earth's global* winds. The Sun heats Earth's surface unevenly because of the ____________ ____________. This uneven heating causes differences in ____________. ⇩ ________ pressure develops over the tropics. ________ pressure develops over the poles. The movement of air from areas of high pressure to areas of low pressure is called ________. ⇩ Global wind belts influence **a.** ________ **b.** ________
Global Winds Belts *I found this on page* ________.	**Assess** *information about circulation in Earth's atmosphere. Read each statement below. If the statement is true, write* true *on the line. If the statement is false, write false on the line and rewrite the underlined portion so that it is true.* Two of the three cells that scientists use to describe circulation of Earth's atmosphere are <u>conduction</u> cells. ________________ The first belt begins with warm air rising at the equator and dropping back to Earth near <u>30° latitude</u>. ________________ The third cell, at the <u>lowest</u> latitude, is also a convection cell. ________________
I found this on page ________.	**Explain** *the Coriolis effect.* Coriolis effect: ________________ ________________ ________________

Main Idea	Details

Analyze *prevailing* winds.

	Winds	Description
I found this on page ________.	Trade	
I found this on page ________.	Doldrums	
I found this on page ________.	Westerlies	
I found this on page ________.	Polar easterlies	
I found this on page ________.	Jet stream	

Local Winds
I found this on page ________.

Compare and contrast *a* sea breeze *and a* land breeze.

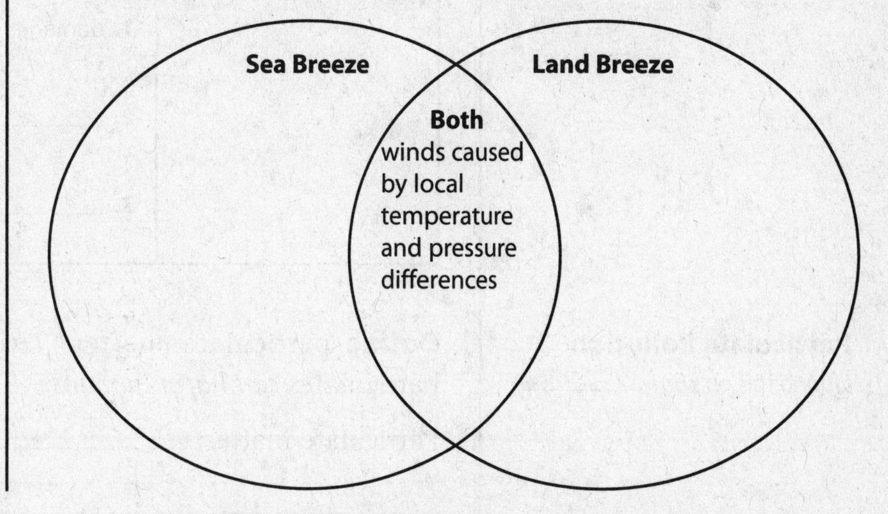

Synthesize It An airplane pilot flying from California to New York would like to make the flight in the shortest amount of time possible. What could the pilot do to decrease his travel time?

____

____

Lesson 4 Air Quality

Scan *Lesson 4. Read the lesson titles and bold words. Look at the pictures. Identify three facts that you discover about Earth's air quality. Write these facts in your Science Journal.*

Main Idea	Details

Sources of Air Pollution

Identify *the 2 general sources of* air pollution, *and give an example of each.*

1. ______________________________

____

2. ______________________________

____

Causes and Effects of Air Pollution

I found this on page ________.

Analyze *the causes and effects of* air pollution.

Cause	Effect
Acid precipitation	1. ____________ 2. ____________
	1. damages plant and animal tissue 2. ____________ 3. ____________

Particulate Pollution

I found this on page ________.

Define particulate matter. *Then list three ways in which* particulates *can harm humans.*

Particulate matter: ______________________________

____

1. ______________________________

2. ______________________________

3. ______________________________

Main Idea	Details
Movement of Air Pollution *I found this on page* ______.	**Identify** *the problems of wind moving or not moving* air pollution. When the wind blows, it ______________________ ______________________. When the wind does not blow, ______________________ ______________________ ______________________.
Maintaining Healthful Air Quality *I found this on page* ______.	**Describe** *two aspects of the Clean Air Act.* **1.** The Clean Air Act gives the U.S. government ______ ______________________. **2.** The standards require states to ______ ______________________.
I found this on page ______.	**Complete** *the statement to explain how monitoring air quality helps people.* **If** air pollution levels are too high . . . **...then** the public is notified of danger and ______________________
Air Quality Trends *I found this on page* ______.	**Identify** *four sources of indoor* air pollution. **1.** ______ **3.** ______ **2.** ______ **4.** ______

Synthesize It Suppose that a doctor has just diagnosed you with a respiratory problem. She has suggested that you remove sources of air pollution from your living space. What could you do?

____

____

____

____

Review Earth's Atmosphere

Chapter Wrap-Up

Now that you have read the chapter, think about what you have learned.

Use this checklist to help you study.

- ☐ Complete your Foldables® Chapter Project.
- ☐ Study your *Science Notebook* on this chapter.
- ☐ Study the definitions of vocabulary words.
- ☐ Reread the chapter, and review the charts, graphs, and illustrations.
- ☐ Review the Understanding Key Concepts at the end of each lesson.
- ☐ Look over the Chapter Review at the end of the chapter.

THE BIG IDEA **Summarize It** Reread the chapter Big Idea and the lesson Key Concepts. Explain how Earth's atmosphere affects life on Earth.

Challenge *What is being done in your community to improve air quality? What could you do to promote the effort?*

Name ______________________________ Date ______________

Weather

How do scientists describe and predict weather?

Before You Read

Before you read the chapter, think about what you know about weather. Record your thoughts in the first column. Pair with a partner, and discuss his or her thoughts. Write those thoughts in the second column. Then record what you both would like to share with the class in the third column.

Think	Pair	Share

Chapter Vocabulary

Lesson 1	Lesson 2	Lesson 3
NEW	**NEW**	**NEW**
weather	low-pressure system	surface report
air pressure	high-pressure system	upper-air report
humidity	air mass	Doppler radar
relative humidity	front	isobar
dew point	tornado	computer model
precipitation	hurricane	
water cycle	blizzard	
REVIEW	**ACADEMIC**	
variable	dominate	
kinetic energy		

Lesson 1 Describing Weather

Scan *Lesson 1. Read the lesson titles and bold words. Look at the pictures. Identify three facts that you discovered about weather. Write these facts in your Science Journal.*

Main Idea	Details
What is weather? *I found this on page* ______.	**Define** weather. Weather: ______
Weather Variables	**Describe** *these variables of* weather.

Variable	Description	How It Is Measured
		thermometer; measured in °C or °F
		barometer; measured in millibars (mb)
		anemometer; measured in mph or km/h
		measured in g/m^3

I found this on page ______.

I found this on page ______.

I found this on page ______.

I found this on page ______.

I found this on page ______.

Examine *which air temperature can hold the greater amount of water vapor. Indicate it by using < or >.*

warm air ◯ cool air

I found this on page ______.

Explain *what a* relative humidity *of 75 percent indicates.*

______

______

Main Idea | Details

I found this on page ________.

Identify *the events that must occur in order for the* dew point *to be reached.*

1. Air temperature ________________________.
2. The amount of moisture in the air ________________.

I found this on page ________.

Sequence *the steps in cloud formation.*

Warm air that contains water vapor ____________ and ________________________.

→ When the cooling air reaches its ________________, ____________ condenses and forms ____________.

→ The ____________ are surrounded by thousands of other ____________. They block and reflect __________, which makes them visible as __________.

Classify *clouds. Describe the appearance of each type of cloud, and identify the altitude at which it is found.*

Type of Cloud	Appearance	Altitude
Stratus		
Cumulus		
Cirrus		

I found this on page ________. (Stratus)

I found this on page ________. (Cumulus)

I found this on page ________. (Cirrus)

I found this on page ________.

Complete *the sentence frame to describe fog.*

Fog is a suspension of ________________ close to Earth's surface.

Main Idea	Details
I found this on page ______.	**Identify** *4 types of* precipitation. *Circle the types that reach Earth's surface as frozen water.* **1.** ______ **3.** ______ **2.** ______ **4.** ______
I found this on page ______.	**Label** *the* water cycle *in the illustration below, and then explain how the* water cycle *relates to* weather.

______

______

______

______

______

Connect It A greenhouse owner determines that the plants in the greenhouse need a higher humidity level. How could the owner address this problem?

______

______

______

______

______

______

Lesson 2 Weather Patterns

Predict *three facts that will be discussed in Lesson 2 after reading the headings. Write these facts in your Science Journal.*

Main Idea	Details
Pressure Systems *I found this on page* ________.	**Compare and contrast** *2 types of pressure systems by completing the Venn diagram. Include a description of the weather that results from each.*

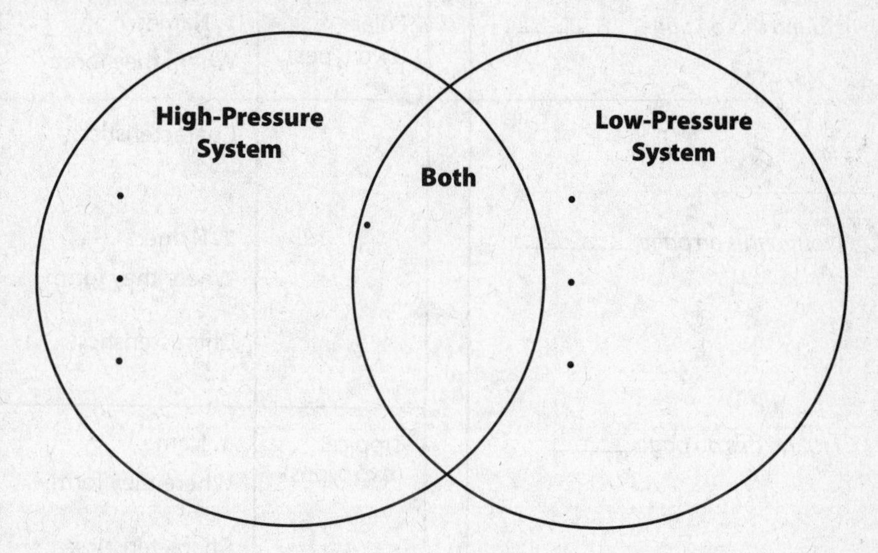

Main Idea	Details
Air Masses *I found this on page* ________.	**Organize** *information about* air masses.

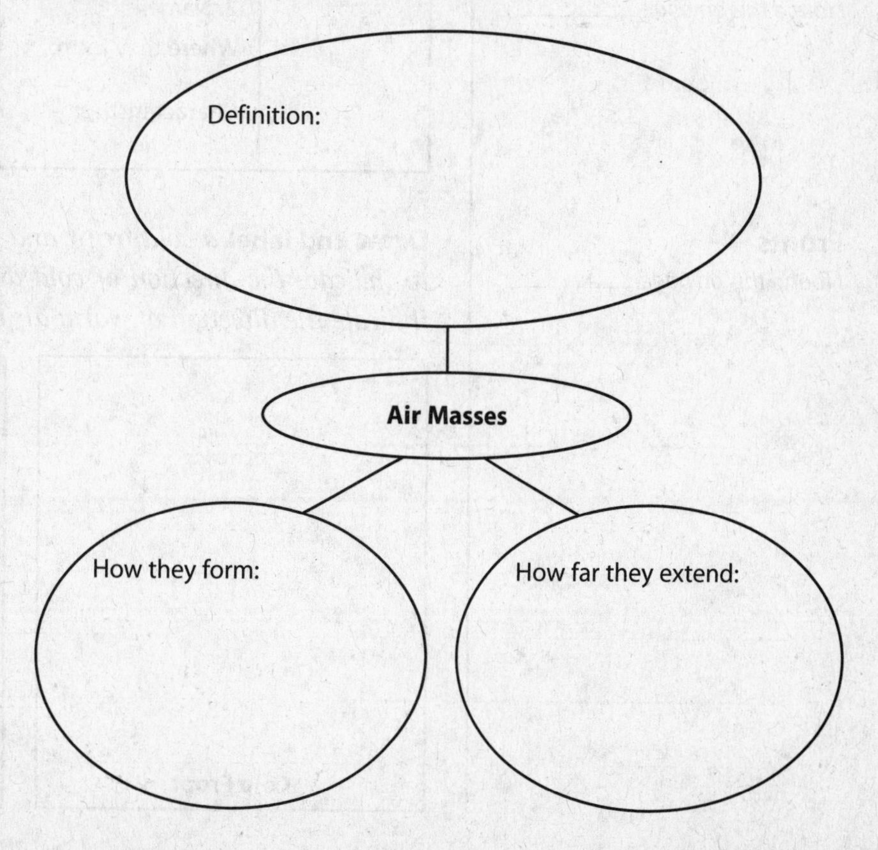

Main Idea | Details

Classify air masses.

Main Idea	Type	Description
I found this on page ______.	Arctic	Where they form: Characteristics:
I found this on page ______. *I found this on page* ______.	Polar (two types)	**1.** Name: Where they form: Characteristics: **2.** Name: Where they form: Characteristics:
I found this on page ______. *I found this on page* ______.	Tropical (two types)	**1.** Name: Where they form: Characteristics: **2.** Name: Where they form: Characteristics:

Fronts

I found this on page ______.

Draw and label *a cold* front *and a warm* front. *Use blue arrows to indicate the direction of cold air movement and red arrows to indicate the direction of warm air movement.*

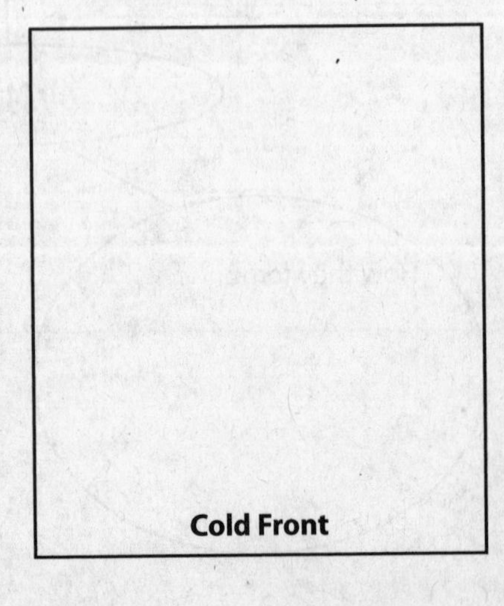

Cold Front

Warm Front

Main Idea	Details
I found this on page ______.	**Define** *stationary and occluded* fronts, *and describe the weather associated with each type.* Stationary front: ______ Occluded front: ______
I found this on page ______.	**Summarize** *why it is useful to understand weather patterns associated with* fronts. ______
Severe Weather *I found this on page* ______.	**Sequence** *the three-stage life cycle of a thunderstorm.* Cumulus stage: ______ Dissipation stage:
I found this on page ______.	**Diagram** *the structure of a* tornado. *Label these parts in your diagram.* • funnel • air inflow • rotating updrafts • air outflow

Main Idea | Details

I found this on page ______.

Sequence *the steps in the formation of a* hurricane.

Warm, moist air ______ and ______. Water vapor ______________, and clouds form. As more air rises, an area of ______________ forms over the ocean.

⇩

As air ________, a ____________________ forms. Air begins to turn ____________________ because of the ____________. Winds are between ____________.

⇩

As air continues to rise and ______, the storm builds to a ______ ______. Winds are greater than ______ but less than ______.

⇩

When winds reach ______, the storm becomes a ______.

I found this on page ______.

Identify *five examples of severe weather.*

1. ______________
2. ______________
3. ______________
4. ______________
5. ______________

I found this on page ______.

Distinguish *weather watches and warnings.*

A ______________ means that severe weather is possible.

A ______________ means that severe weather is already occurring.

Analyze It Town A experiences several days of cold temperatures and steady rain. Town B, which is twenty kilometers east of Town A, experiences rain and warm temperatures during that same time. What weather pattern explains these events?

____

____

____

____

Lesson 3 Weather Forecasts

Skim *Lesson 3 in your book. Read the headings and look at the photos and illustrations. Identify three things you want to learn more about as you read the lesson. Record your ideas in your Science Journal.*

Main Idea — Details

Measuring the Weather
I found this on page ________.

Describe *the first step in making a weather forecast, and identify three instruments used to measure weather variables.*

Step 1: Measure the condition of the ________________ using weather instruments, such as

a. a ______________________, which measures air temperature,

b. a barometer, which measures ______________________, and

c. an ______________________, which measures wind speed.

I found this on page ________.

Identify *how scientists measure weather conditions in different parts of the atmosphere.*

Measuring Conditions of the Atmosphere

Surface Reports	Upper-Air Reports

I found this on page ________.

Compare *2 types of satellite images.*

Visible Light Image	Infrared Image

I found this on page ________.

Organize *information about* Doppler radar.

Doppler is a specialized type of ________.

Doppler radar can detect	Doppler radar can estimate
________________.	________________.

Main Idea / Details

Weather Maps

I found this on page ________.

Identify *the types of information displayed on a station model.*

1. ____________________
2. ____________________
3. ____________________
4. ____________________
5. ____________________
6. __________ and __________

I found this on page ________.

Contrast isobars *and* isotherms.

Isobars	Isotherms

I found this on page ________.

Identify *each symbol found on weather maps.*

Symbol	Meaning
* *	
••	
H	
L	

Main Idea | Details

I found this on page ________.

Analyze *the weather map. Color a high-pressure area red. Color a warm front yellow. Color an occluded front blue.*

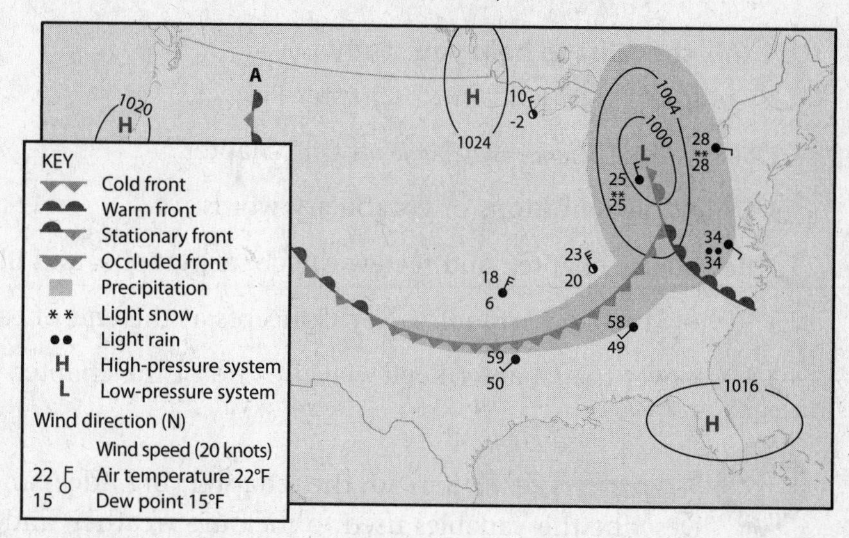

Predicting the Weather

I found this on page ________.

Sequence *how weather* computer models *are generated and distributed.*

Government offices exchange information ________________.

→ Computer model programs solve

→ The formulas predict

→ Weather maps and forecasts are made available through

Synthesize It Which type of map would better help you plan next weekend's activities, a station map or a weather map? Explain why.

____

____

____

____

____

Review Weather

Chapter Wrap-Up

Now that you have read the chapter, think about what you have learned.

Use this checklist to help you study.

- ❑ Complete your Foldables® Chapter Project.
- ❑ Study your *Science Notebook* on this chapter.
- ❑ Study the definitions of vocabulary words.
- ❑ Reread the chapter, and review the charts, graphs, and illustrations.
- ❑ Review the Understanding Key Concepts at the end of each lesson.
- ❑ Look over the Chapter Review at the end of the chapter.

THE BIG IDEA **Summarize It** Reread the chapter Big Idea and the lesson Key Concepts. Describe the variables used to measure weather, and explain how weather systems can be used to predict future weather.

Challenge *Design a weather station for your school. What instruments will you use? How will you record the collected data? How could you make the information available to others?*

Name ____________________ Date ____________

Climate

What is climate and how does it impact life on Earth?

Before You Read

Before you read the chapter, think about what you know about climate. In the first column, record three things you already know about climate. In the second column, write down three things you would like to learn about climate. When you have completed the chapter, think about what you have learned and complete the **What I Learned** *column.*

K What I Know	W What I want to Learn	L What I Learned

Chapter Vocabulary

Lesson 1	Lesson 2	Lesson 3
NEW climate rain shadow specific heat microclimate **REVIEW** precipitation	**NEW** ice age interglacial El Niño/Southern Oscillation monsoon drought **ACADEMIC** phenomenon	**NEW** global warming greenhouse gas deforestation global climate model

Lesson 1 Climates of Earth

Scan *Lesson 1. Record three questions you have about Earth's climates in your Science Journal. Try to answer your questions as you read.*

Main Idea / Details

What is climate?
I found this on page ________.

Define climate.

Climate: ____

____

What affects climate?
I found this on page ________.

Identify *4 factors that determine a region's* climate.

1. ____

2. ____

3. ____

4. ____

I found this on page ________.

Describe *how latitude affects* climate *in different regions on Earth.*

Polar regions: ____

____

Locations near the equator: ________________________________

____

Middle latitudes: ____

____

I found this on page ________.

Compare *how altitude and latitude influence temperature.*

Altitude	Latitude
As altitude increases, temperature ________.	As latitude increases, temperature ________.

Main Idea	Details
I found this on page ______.	**Sequence** *the events that result in a* rain shadow.

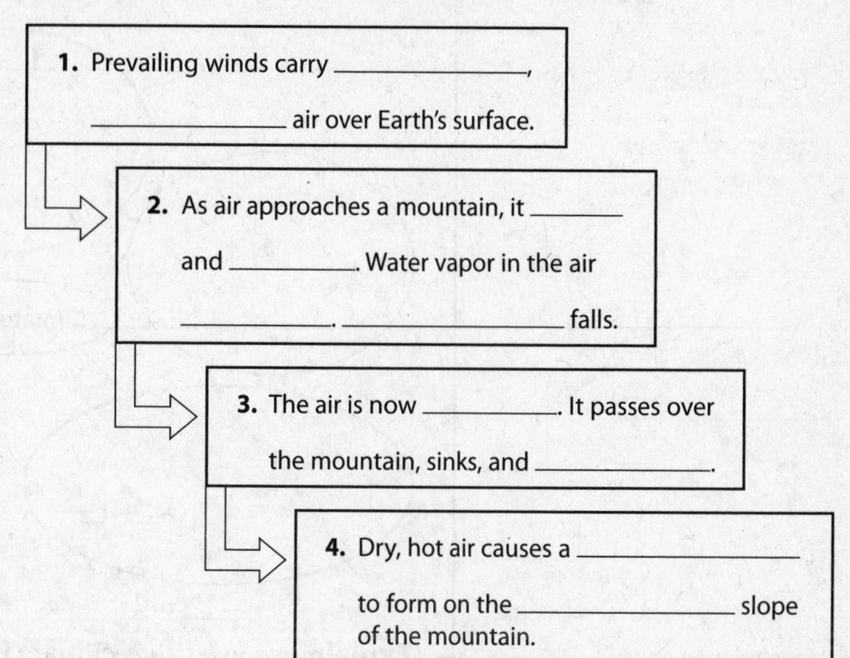

Large Bodies of Water
I found this on page ______.

Define specific heat, *and then explain how the* specific heat *of water can influence the* climate *of an area.*

Specific heat: ____________________

____

Explanation: ____________________

____

____

____

____

Classifying Climates
I found this on page ______.

Define microclimate, *and identify three* microclimates.

Microclimate: ____________________

____

____

1. ____________________

2. ____________________

3. ____________________

Main Idea	Details
I found this on page ______.	**Identify and describe** *Koppen's 5 climate types.*

Climate Types

Main Idea	Details
I found this on page ______.	**Explain** *two ways that* climate *can affect people.*
	Agriculture: ______
	Architecture: ______
I found this on page ______.	**Analyze** *how* climate *affects each of the following organisms.*
	Polar bears: ______
	Desert plants and animals: ______
	Deciduous trees: ______

Connect It Classify the climate in your area, and give reasons for your classification.

______

______

______

Lesson 2 Climate Cycles

Predict *three facts that will be discussed in Lesson 2 after reading the headings. Record your facts in your Science Journal.*

Main Idea	Details
Long-Term Cycles *I found this on page* ______.	**Distinguish** *four ways scientists learn about past climates.* **1.** ______ **2.** ______ **3.** ______ **4.** ______
I found this on page ______.	**Compare** *an* ice age *with an* interglacial. Ice age: ______ ______ Interglacial: ______ ______
I found this on page ______.	**Model** *the time spanned by Earth's most recent* ice age *and* interglacial *on the time line. Use these labels:* • Ice age begins • Maximum ice coverage • Holocene interglacial begins
I found this on page ______.	**Identify** *four causes of Earth's long-term climate cycles.* **1.** ______ **2.** ______ **3.** ______ **4.** ______

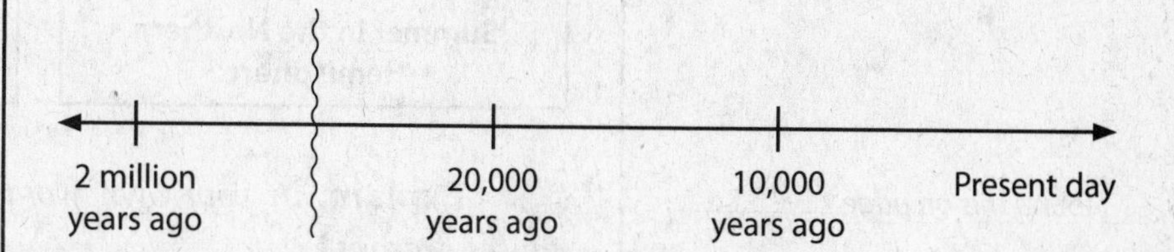

Main Idea	Details
Short-Term Cycles *I found this on page* ______.	**Summarize** *two causes of short-term climate cycles.*

Causes of short-term climate cycles

- (blank)
- interaction between

I found this on page ______.

Diagram *the position of Earth and its axis in relation to the Sun during summer and winter in the northern hemisphere.*

Summer in the Northern Hemisphere	Winter in the Northern Hemisphere

I found this on page ______.

Explain, *in your own words, how the tilt of Earth's axis causes seasons.*

Main Idea	Details

I found this on page ______.

Review *how Earth's equinoxes and solstices mark the beginning of each of the 4 seasons in this organizer.*

The Beginnings of Seasons

Solstices	Equinoxes
• Summer: axis ______ to the Sun • Winter: axis ______ from the Sun	• axis tilted so that both northern and southern hemispheres receive ______ amounts of sunlight. • beginning of ______ and ______

I found this on page ______.

Sequence *the statements to describe the phenomenon of* El Niño/Southern Oscillation.

____ Warm water surges back to South America, preventing cold water from upwelling.

____ Trade winds that blow from east to west weaken.

____ The normal pattern of high and low pressure across the Pacific is reversed.

I found this on page ______.

Compare ENSO *and* NAO *weather patterns.*

ENSO: El Nino/Southern Oscillation	NAO: North Atlantic Oscillation
Description:	Description:
Weather Pattern:	Weather Pattern:

Main Idea	Details
I found this on page ______.	**Explain** *how* monsoons *change with the seasons.* Summer: ______ ______ ______ ______ Winter: ______ ______ ______
Droughts, Heat Waves, and Cold Waves *I found this on page* ______.	**Define** drought. Drought: ______
I found this on page ______.	**Model** *the results of a* drought *and a heat wave occurring at the same time.* **Drought** **Heat Wave** **a.** ______ damage **b.** ______ shortages **c.** loss of ______
I found this on page ______.	**Describe** *the cause of cold waves.* ______ ______ ______

Analyze It Review what can happen during a drought and heat wave. What might be the effect of a cold wave?

______

______

______

______

Lesson 3 Recent Climate Change

Skim *Lesson 3 in your book. Read the headings and look at the photos and illustrations. Identify three things that you want to learn more about. Write your ideas in your Science Journal.*

Main Idea	Details
Regional and Global Climate Change *I found this on page* ______.	**Summarize** *Earth's air temperature over the past 100 years.* 1880–1900 temperatures ______ 1900–1945 temperatures ______ 1945–1975 temperatures ______ 1975–2000+ temperatures ______
Human Impact on Climate Change *I found this on page* ______.	**Define** global warming, *and explain the conclusion of the Intergovernmental Panel on Climate Change (IPCC).* Global warming: ______ ______ The IPCC concluded that most of the temperature increase is due to ______ such as burning ______ and ______.
I found this on page ______.	**Explain** *how* greenhouse gases *affect Earth's temperatures.*

Natural Greenhouse Effect

______ (CO_2), ______, and water vapor ______ Earth's outgoing infrared ______.	⇨	Temperatures suitable for life are ______.

Over the Last 120 Years

An increase in CO_2 levels causes ______ ______.	⇨	Average surface temperatures have been ______.

Main Idea	Details
I found this on page ______.	**Identify** *three natural sources of carbon dioxide (CO_2).* 1. ______ 2. ______ 3. ______
I found this on page ______.	**Recall** *two human-caused sources of carbon dioxide (CO_2).* Human sources → ______ → ______
I found this on page ______.	**Explain** *how aerosols are released into the atmosphere.* Burning → releases aerosols, which are → Aerosols ______ sunlight back into space, which results in ______ temperatures. → Clouds with ______ amounts of aerosols have ______ cloud droplets that ______ more sunlight, which ______ climate.
Climate and Society *I found this on page* ______.	**Identify** *the problems climate change poses for society.*

Cause →	Effect
______ and ______	food and water shortages
excessive rainfall	

Main Idea	Details
I found this on page ______.	**Explain** *the environmental impacts of climate change.*

Warmer Temperatures
a. cause more water to ______, producing ______ and ______ storms.
b. ______ glaciers and polar ice sheets and cause ______ to rise.
c. melt the frozen ______ in the Arctic, changing ______ patterns.
d. cause ______ weather events to become more common.

Predicting Climate Change
I found this on page ______.

Define *the* global climate model (GCM), *and explain the limitations of the model's predictions.*

GCM: ______

I found this on page ______.

Describe *two activities of increasing human populations that might affect climate.*

1. ______

2. ______

I found this on page ______.

Identify *two ways people can reduce* greenhouse gases.

1. ______

2. ______

Synthesize It Identify ways that people in your community could help to reduce greenhouses gases.

______

Review **Climate**

Chapter Wrap-Up

Now that you have read the chapter, think about what you have learned. Complete the **What I Learned** *column on the first page of the chapter.*

Use this checklist to help you study.

- ❑ Complete your Foldables® Chapter Project.
- ❑ Study your *Science Notebook* on this chapter.
- ❑ Study the definitions of vocabulary words.
- ❑ Reread the chapter, and review the charts, graphs, and illustrations.
- ❑ Review the Understanding Key Concepts at the end of each lesson.
- ❑ Look over the Chapter Review at the end of the chapter.

THE BIG IDEA **Summarize It** Reread the chapter Big Idea and the lesson Key Concepts. Analyze the information you have learned about climate. Explain how climate affects your life.

____

____

____

____

____

____

____

____

____

____

____

____

____

Challenge *Examine the effects of urban sprawl in your area. Describe how the changes made by an expanding city might affect the local climate.*

Name ______________________________ Date ____________

Motion, Forces, and Newton's Laws

In what ways do forces affect an object's motion?

Before You Read

Before you read the chapter, think about what you know about motion and forces. Record three things that you already know about motion and forces in the first column. Then write three things that you would like to learn about in the second column. Complete the final column of the chart when you have finished the chapter.

K What I Know	W What I Want to Learn	L What I Learned

Chapter Vocabulary

Lesson 1	Lesson 2	Lesson 3
NEW motion reference point distance displacement speed velocity acceleration **ACADEMIC** satellite	**NEW** force contact force noncontact force friction gravity balanced forces unbalanced forces	**NEW** inertia Newton's first law of motion Newton's second law of motion Newton's third law of motion force pair

Lesson 1 Describing Motion

Scan *Lesson 1. Read the lesson titles and bold words. Look at the pictures. Identify three facts you discovered about motion. Record your facts in your Science Journal.*

Main Idea | Details

Motion
I found this on page __________.

Describe *an example of* motion. *Include an object, a* reference point, distance, *and* direction.

Example:	
Object:	Direction:
Reference point:	Distance:

I found this on page __________.

Differentiate *between* distance *and* displacement *for an object that traveled from point A to point B as shown.*

A B

Distance:

__________________ units

Displacement:

__________________ units

Speed
I found this on page __________.

Define speed.

____

____

Velocity
I found this on page __________.

Characterize velocity *as represented by these two arrows.*

A

B

Feature	What It Means
Direction	
Length of arrows	
Segments of arrows	

Main Idea | Details

I found this on page ______.

Explain *why each example does or does not represent* acceleration. *Circle the examples that do describe* acceleration.

Example	Explanation
A rock falling from a cliff moves faster and faster as it approaches the ground.	
A book sits on a shelf in a classroom.	
A butterfly clings to the paddle of a garden windmill in a steady breeze.	
A motorboat moves north across the lake at 20km/h.	
An arrow shot from a bow arcs high into the air and then plunges into a bail of hay on the ground.	

Calculating Acceleration

I found this on page ______.

Point out *parts of the formula for average* acceleration.

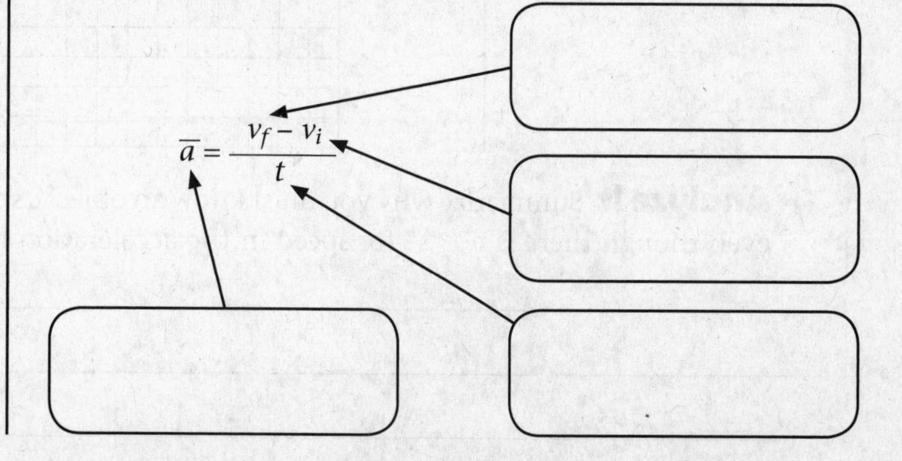

Main Idea	Details
I found this on page ______.	**Differentiate** *positive and negative* acceleration.
Using Graphs to Represent Motion I found this on page ______.	**Diagram** motion. **A.** *Draw a displacement-time graph. Label the axes, and represent an animal that:* • *starts speeding up right away.* • *rests, and then* • *speeds away from the resting point.*
I found this on page ______.	**B.** *Draw a speed-time graph. Label the axes. Represent an animal that:* • *remains at rest for the three hours,* • *steadily increases speed for three hours, and then* • *travels at a constant rate for three hours.*

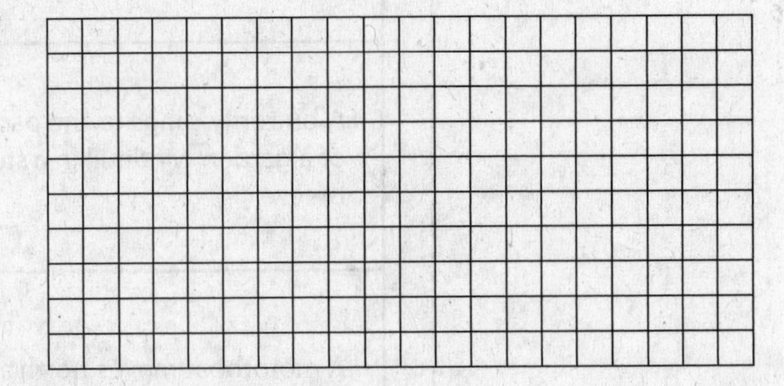

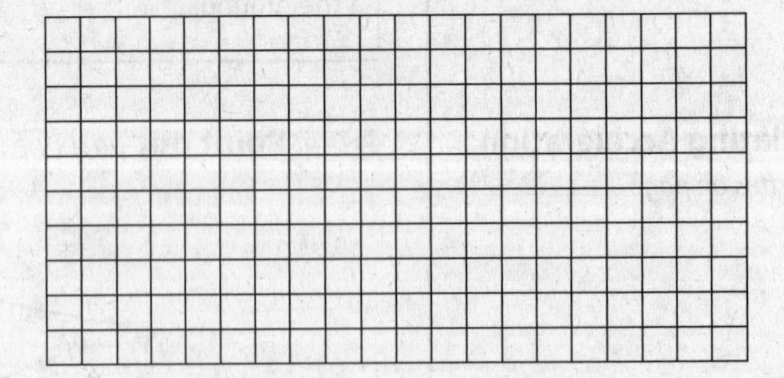

Analyze It Summarize why you must know an object's speed to calculate acceleration even though there is no "s" for speed in the acceleration formula.

Lesson 2 Forces

Predict *three facts that will be discussed in Lesson 2 after reading the headings. Record your predictions in your Science Journal.*

Main Idea | Details

What are forces?
I found this on page ________.

Characterize forces.

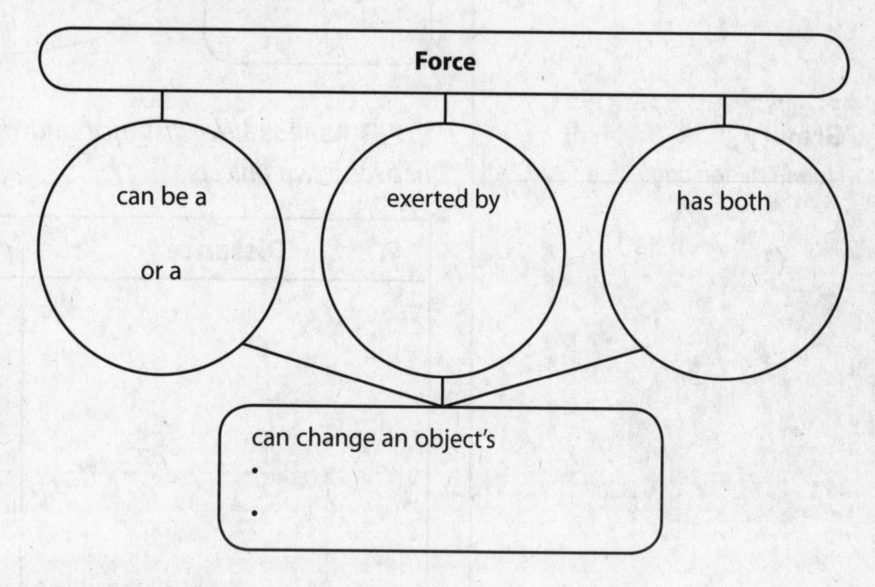

Types of Forces
I found this on page ________.

Classify forces. *Differentiate* contact forces *from* noncontact forces. *Circle the* contact forces *in red. Circle the* noncontact forces *in blue.*

Type of Force	Example
Applied	
Elastic	
Normal	
Electric	
Magnetic	
Gravity	

Main Idea	Details

Friction
I found this on page ________.

Identify *factors that affect the* force *of* friction.

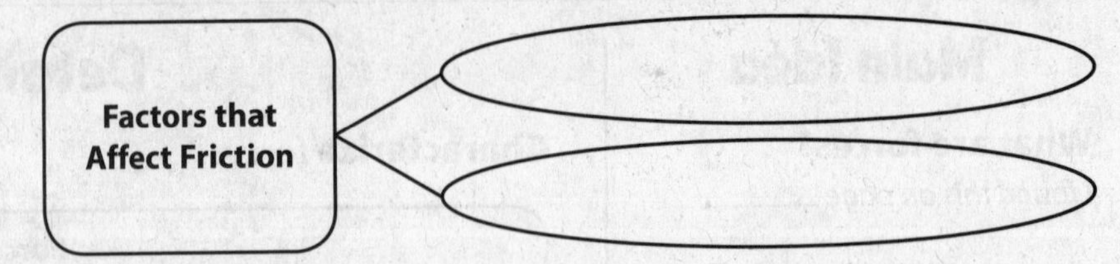

Gravity
I found this on page ________.

Assess *how distance and mass affect the* force *of* gravity *between two objects.*

Distance	Mass

Combining Forces
I found this on page ________.

Calculate *net* forces. *Circle* balanced forces *in red and* unbalanced forces *in blue.*

Combined Forces	Net Force (Draw an arrow to show direction.)
→ 30 N → 70 N	
→ 30 N ← 40 N	
→ 60 N ← 60 N	
→ 18 N → 12 N ← 30 N	

Synthesize It Suppose that you want to build a machine to perform some task. Why must you understand all about forces to complete your mission?

____

____

____

Lesson 3 Newton's Laws of Motion

Skim *Lesson 3 in your book. Read the headings and look at the photos and illustrations. Identify three things you want to learn more about as you read the lesson. Record your ideas in your Science Journal.*

Main Idea	Details
Newton's Laws *I found this on page* ______.	**Relate** *details about Isaac Newton.*

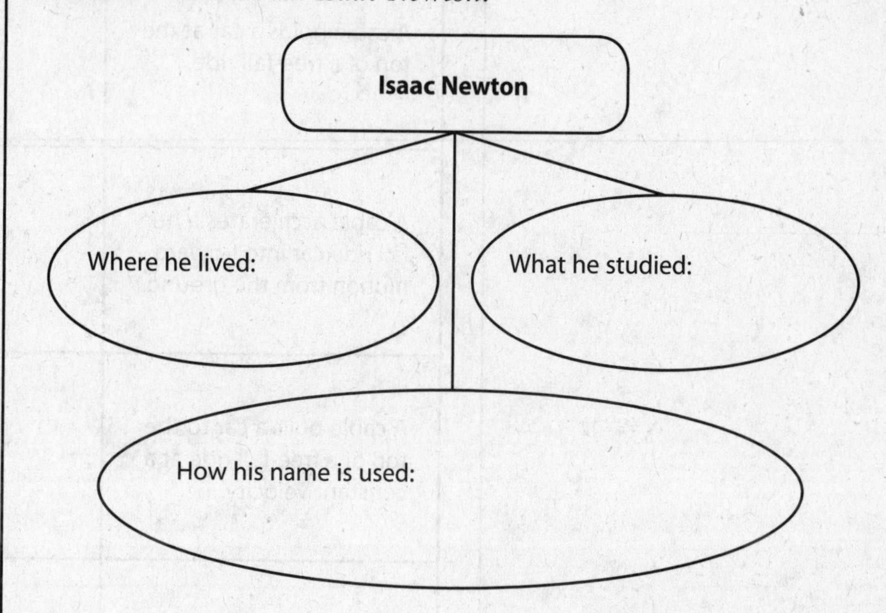

Main Idea	Details
Newton's First Law *I found this on page* ______.	**Define** inertia. ______ ______
I found this on page ______.	**Diagram** *the concept of* Newton's first law of motion.

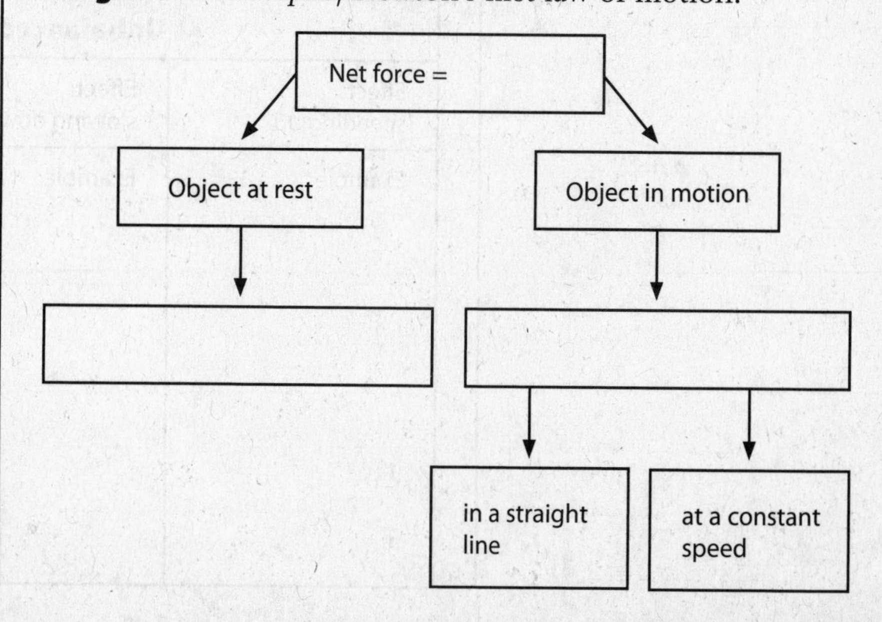

Main Idea | Details

I found this on page ________.

Describe *the forces in each scenario. Circle the examples of balanced forces in red and unbalanced forces in blue.*

Scenario	Description of Forces
A cable holds a car at the top of a free-fall ride.	
A cable accelerates a free-fall ride car into upward motion from the ground.	
A cable pulls a car to the top of a free-fall ride at a constant velocity.	
A cable releases; a free-fall ride car accelerates to the ground.	

I found this on page ________.

 **Explain** *three effects of unbalanced forces.*

Unbalanced Forces		
Effect: speeding up	Effect: slowing down	Effect: changing direction
Example:	Example:	Example:

Main Idea	Details
Newton's Second Law of Motion *I found this on page* ______.	**Identify** *the parts of the formula described by* Newton's second law of motion.

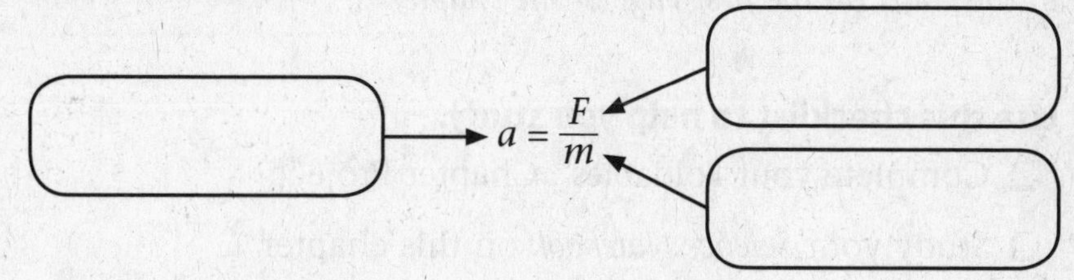

I found this on page ______.

Generalize *the relationship between mass, acceleration, and net force.*

______

______

Newton's Third Law
I found this on page ______.

Paraphrase Newton's third law of motion.

______

______

I found this on page ______.

Organize *information about* force pairs.

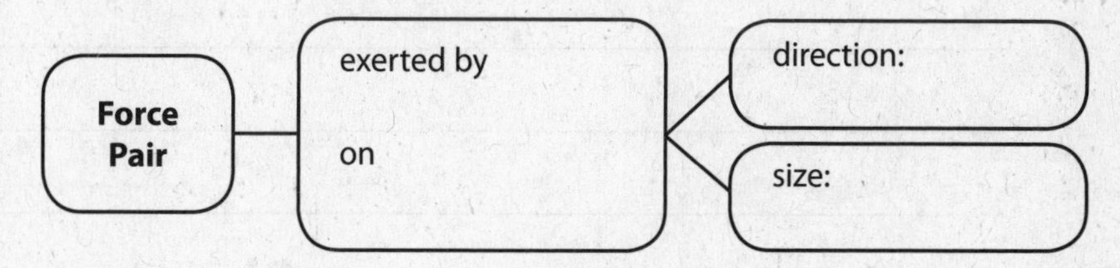

Newton's Laws in Action
I found this on page ______.

Explain *two situations in which Newton's laws do not apply to the motion of objects.*

1. ______

2. ______

Connect It Describe how Newton's first, second, and third laws apply to you eating your breakfast.

______

______

______

______

______

Review

Motion, Forces, and Newton's Laws

Chapter Wrap-Up

Now that you have read the chapter, think about what you have learned. Complete the final column in the chart on the first page of the chapter.

Use this checklist to help you study.

- ❑ Complete your Foldables® Chapter Project.
- ❑ Study your *Science Notebook* on this chapter.
- ❑ Study the definitions of vocabulary words.
- ❑ Reread the chapter, and review the charts, graphs, and illustrations.
- ❑ Review the Understanding Key Concepts at the end of each lesson.
- ❑ Look over the Chapter Review at the end of the chapter.

THE BIG IDEA **Summarize It** Reread the chapter Big Idea and the lesson Key Concepts. Summarize why games and fun activities make good examples to explain the principles of motion and forces.

____

____

____

____

____

____

____

____

____

____

____

____

____

Challenge *Choose your favorite sport. Do an analysis of the forces and motion that occur in a typical game. Write and illustrate a descriptive essay that summarizes your analysis, and share your essay with your class.*

Name ______________________________ Date ____________

The Sun-Earth-Moon System

What natural phenomena do the motions of Earth and the Moon produce?

Before You Read

Before you read the chapter, think about what you know about phenomena caused by the motions of Earth and the Moon. Record your thoughts in the first column. Pair with a partner, and discuss his or her thoughts. Write those thoughts in the second column. Then record what you both would like to share with the class in the third column.

Think	Pair	Share

Chapter Vocabulary

Lesson 1	Lesson 2	Lesson 3
NEW orbit revolution rotation rotation axis solstice equinox **ACADEMIC** equator	**NEW** maria phase waxing phase waning phase	**NEW** umbra penumbra solar eclipse lunar eclipse tide

Lesson 1 Earth's Motion

Skim *Lesson 1 in your book. Read the headings and look at the photos and illustrations. Write three things you want to learn more about as you read the lesson. Write your ideas in your Science Journal.*

Main Idea / Details

Earth and the Sun

I found this on page ________.

Organize *information about the Sun.*

The Sun
- about ________ km from Earth
- energy from ________ ________
- core temperature: more than ________ surface temperature: ________
- Sun's energy reaches Earth as ________ and ________

I found this on page ________.

Complete *information about Earth's* revolution *around the Sun.*

Earth Revolves Around the Sun	
Definition of revolution	
How long it takes Earth to make one revolution around the Sun	
Definition of orbit	
Force that keeps Earth in its orbit around the Sun	

Main Idea	Details

I found this on page ______.

Arrange *facts about Earth's* rotation.

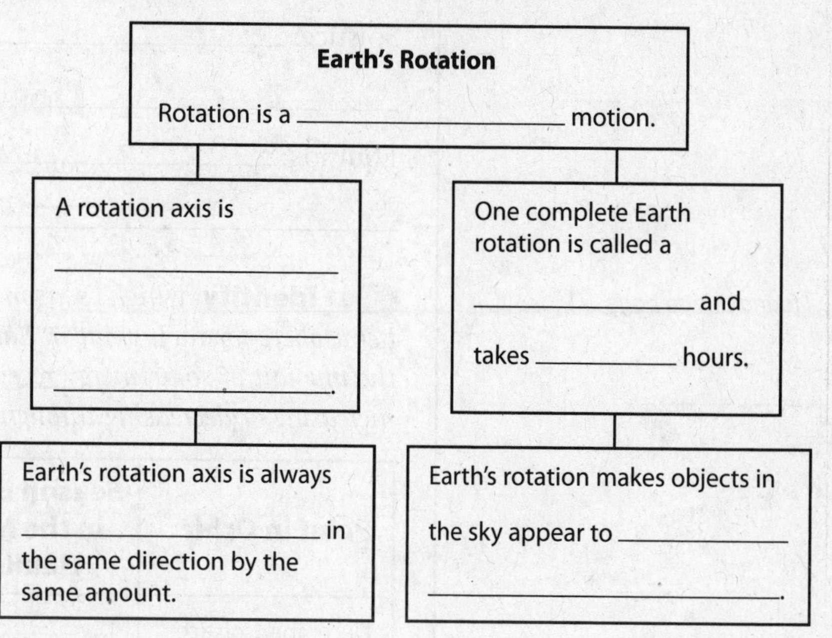

Temperature and Latitude

I found this on page ______.

Analyze *the interaction of sunlight with Earth's surface.*

Cause	Effect
Curved surface of Earth	The energy in a beam of sunlight is spread out more at ______ than at ______. This makes Earth ______ at the poles and ______ at the equator.

Seasons

I found this on page ______.

Identify *the direction of the north end of Earth's* rotation axis *for each of the four seasons. Put a check mark in the appropriate column.*

Direction in which Earth's rotation axis is leaning			
Season	**Toward the Sun**	**Away from the Sun**	**Neither toward nor away**
Winter			
Spring			
Summer			
Fall			

Main Idea	Details

I found this on page ______.

Define solstice *and* equinox.

Solstice: ______________________________

____

Equinox: ______________________________

____

I found this on page ______.

Identify *which season is beginning in the northern hemisphere for each point in Earth's* orbit. *Then indicate whether the amount of solar energy received by the northern hemisphere is increasing or decreasing throughout each season.*

Point in Orbit	Season Beginning in the Northern Hemisphere	Change in Solar Energy Received
December solstice		
March equinox		
June solstice		
September equinox		

I found this on page ______.

Describe *the height of the apparent path of the Sun through the sky in the northern hemisphere at each* solstice.

December solstice: ______________________________

June solstice: ______________________________

Synthesize It Suppose that Earth's axis were tilted 90 degrees instead of 23.5 degrees. What might the seasons be like?

____

____

____

____

____

____

____

Earth's Moon

Scan *Lesson 2 in your book. In your Science Journal, write three questions you have about the Moon. Try to answer your questions as you read.*

Main Idea / Details

Seeing the Moon
I found this on page ______.

Explain *why you can see the Moon.*

____

____

The Moon's Formation
I found this on page ______.

Sequence *events in the Moon's formation.*

1. A collision between ______________________ ______________________ occurred.

2. ______________________ was ejected into space, and a ring formed ______________________.

3. Material in the ring ______________________ ______________________ and formed ______________________.

I found this on page ______.

Describe *three features of the Moon's surface.*

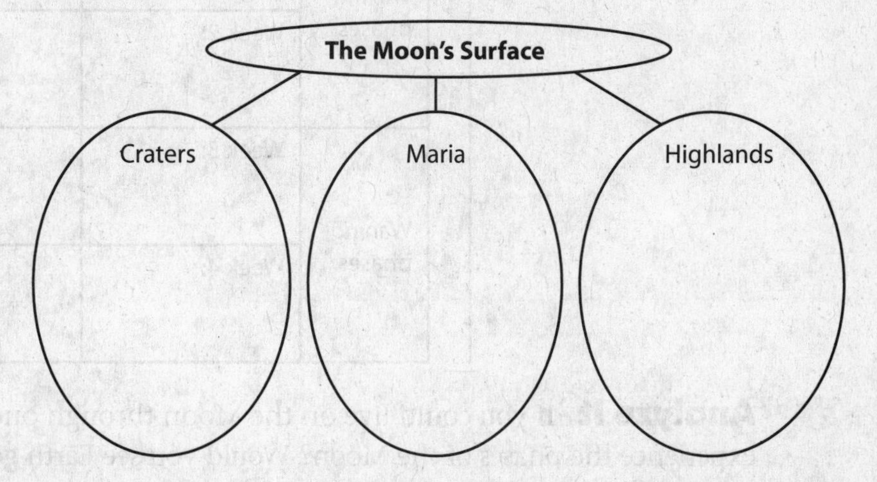

The Moon's Motion
I found this on page ______.

Identify *details about the 2 motions of the Moon.*

	Revolution	Rotation
Period in days		
Description of motion		

Main Idea	Details
I found this on page ______.	**Explain** *why the same side of the Moon always faces Earth.*

____

____

Phases of the Moon

I found this on page ______.

Organize *information about the Moon's* phases.

Phases of the Moon

- Definition:
- Caused by:
- Length of a complete lunar cycle:

I found this on page ______.

Categorize *information about the* phases *of the Moon.*

Phase	Name	Description
Waxing phases	Week 1: First Quarter	
	Week 2:	
Waning phases	Week 3:	
	Week 4:	

Analyze It If you could live on the Moon through one lunar cycle, how would you experience the phases of the Moon? Would you see Earth going through phases? Explain.

____

____

____

____

____

Lesson 3 Eclipses and Tides

Predict *three things you will learn about in Lesson 3. Look at the illustrations in the lesson to give you some clues. Write your predictions in your Science Journal.*

Main Idea	Details

Shadows—the Umbra and the Penumbra
I found this on page ________.

Define umbra *and* penumbra. *Then label the* umbra *and the* penumbra *on the diagram below.*

Umbra: ____

____

Penumbra: ______________________________________

____

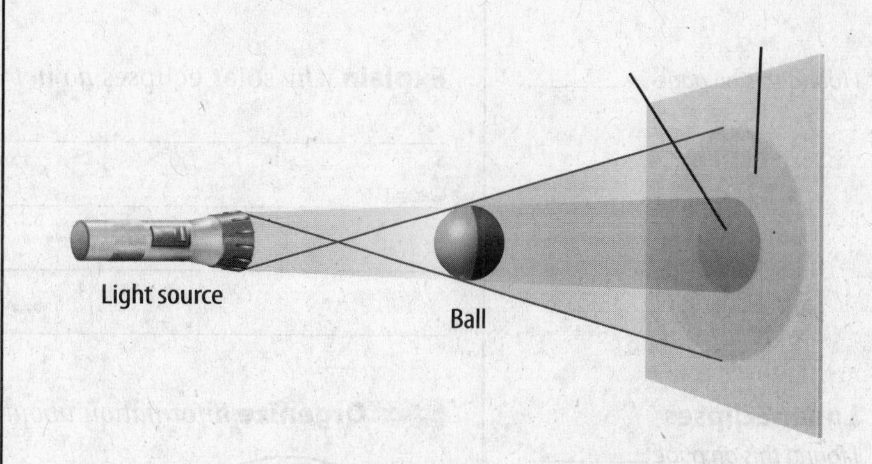

Solar Eclipses
I found this on page ________.

Compare *information about* solar eclipses.

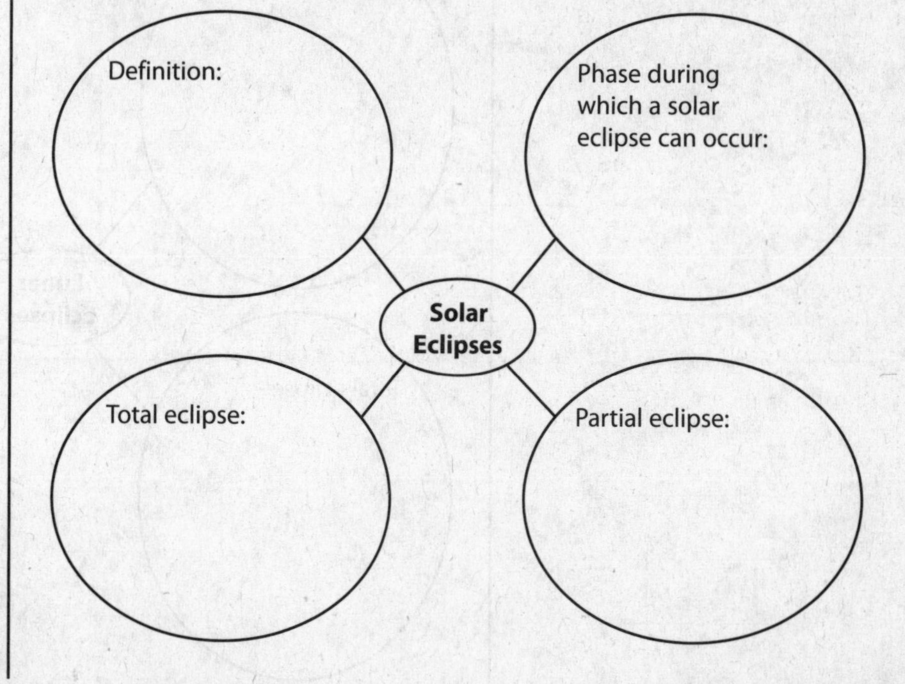

Main Idea	Details
I found this on page ________.	**Label** *the diagram of a* solar eclipse. *Use these terms:* • Sun • Earth • Moon • umbra • penumbra • total solar eclipse • partial solar eclipse

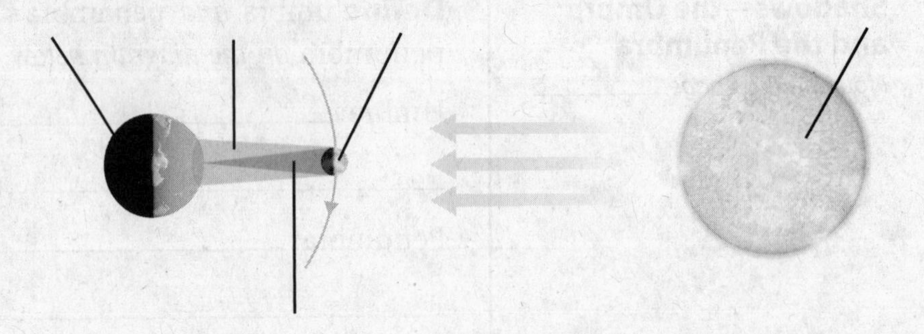

Main Idea	Details
I found this on page ________.	**Explain** *why* solar eclipses *do not occur every month.*
Lunar Eclipses *I found this on page* ________.	**Organize** *information about* lunar eclipses.

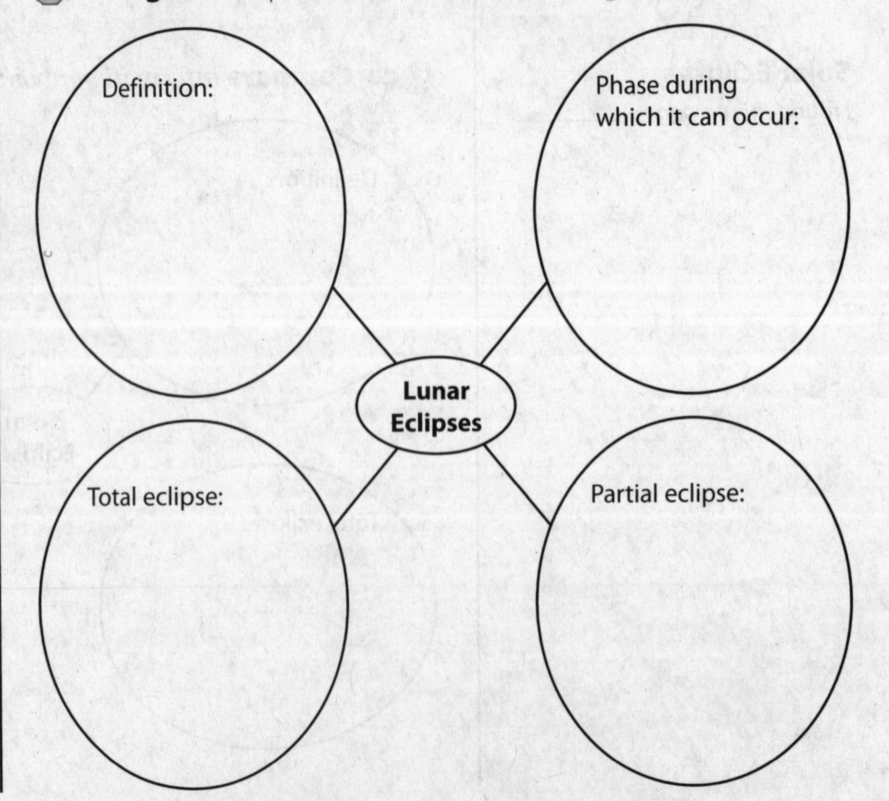

Main Idea | Details

I found this on page ________.

Explain *why you can still see the Moon during a total* lunar eclipse.

____

____

Tides

I found this on page ________.

Organize *information about* tides.

Tides	
Definition	
Force that causes tides	
Where low tide occurs	
Where high tide occurs	
How often high tide occurs	

I found this on page ________.

Compare *spring* tides *and neap* tides. *Draw the position of the Moon, the Sun, and Earth during a spring* tide *and a neap* tide.

Types of Tides	
Spring	**Neap**
Moon phases during which they occur:	Moon phases during which they occur:

Analyze It Suppose that the Moon were smaller in size but greater in mass than it is now. How would that affect solar eclipses and tides?

____

____

____

____

Review The Sun-Earth-Moon System

Chapter Wrap-Up

Now that you have read the chapter, think about what you have learned.

Use this checklist to help you study.

- ❑ Complete your Foldables® Chapter Project.
- ❑ Study your *Science Notebook* on this chapter.
- ❑ Study the definitions of vocabulary words.
- ❑ Reread the chapter, and review the charts, graphs, and illustrations.
- ❑ Review the Understanding Key Concepts at the end of each lesson.
- ❑ Look over the Chapter Review at the end of the chapter.

Summarize It Reread the chapter Big Idea and the lesson Key Concepts. Draw the Earth-Moon-Sun system in the space below. Use arrows to show the rotation and revolution of Earth and the Moon. On the lines below your drawing, describe a cycle that is caused by the motions of Earth and the Moon.

Challenge *Work with a partner to create a moving model of the Sun-Earth-Moon system. Use your model to demonstrate tides, phases of the Moon, seasons, and other cycles caused by motions of Earth around the Sun and the Moon around Earth.*

Name ______________________________ Date __________

Exploring Our Solar System

How and where do scientists look for life in the solar system?

Before You Read

Before you read the chapter, think about what you know about the solar system. Record your ideas in the first column. Pair with a partner, and discuss his or her thoughts. Write those thoughts in the second column. Then record what you both would like to share with the class in the third column.

Think	Pair	Share

Chapter Vocabulary

Lesson 1	Lesson 2	Lesson 3
NEW astronomical unit planet dwarf planet satellite meteoroid meteor meteorite **REVIEW** silicates	**NEW** astrobiology organic geyser	**NEW** artificial satellite rocket space probe **ACADEMIC** transmit

Lesson 1 Our Solar System

Skim *Lesson 1 in your book. Read the headings and look at the photos and illustrations. Identify three things you want to learn more about as you read the lesson. Record your ideas in your Science Journal.*

Main Idea	Details

Origin and Structure of Our Solar System

I found this on page ______.

Sequence *the formation of the solar system. Use the words in parentheses in your explanations.*

spinning cloud of hydrogen gas and dust in space

↓

(gravity)

↓

(nuclear, Sun)

↓

(flattened)

↓

(planetesimals)

↓

(planets, asteroids)

I found this on page ______.

Record *the distance of an* astronomical unit *(AU), and tell what this unit represents.*

1 AU = approximately ______

Represents: ______

I found this on page ______.

Classify *regions of the solar system.*

	Region		
	Inner Solar System	Outer Solar System	Oort Cloud
Contents	• • • • •	• • • •	•

Main Idea | Details

Movement in the Solar System

I found this on page ______.

Categorize *the direction of motion of selected objects within the solar system.*

Object	Revolution	Rotation
Sun	stationary	
Venus		
Earth		
Mars		
Uranus		

Objects in Our Solar System

I found this on page ______.

Relate *conditions that determined formation of different objects in the solar system.*

Object	Star	Rocks and metals	Ices
Location			
Conditions			

I found this on page ______.

Characterize planets *in the solar system.*

Characteristics:
-
-
-

Planet

Examples:
-
-
-
-
-
-
-
-

Main Idea	Details

I found this on page ________.

Characterize dwarf planets.

Dwarf Planets

Characteristics:
•
•
•

Examples:
•
•
•

I found this on page ________.

Define satellite.

____

I found this on page ________.

Compare and contrast *the asteroid belt with the Kuiper belt.*

Asteroid Belt	Both	Kuiper Belt

I found this on page ________.

Differentiate meteoroids, meteors, *and* meteorites.

Meteoroid	Meteor	Meteorite

Synthesize It Explain how the location of objects, relative to the center of the solar system, relates to their formation.

____

____

Lesson 2 Life in the Solar System

Scan *Lesson 2. Read the lesson titles and bold words. Look at the pictures. Identify three facts you discovered about life in the solar system. Record your facts in your Science Journal.*

Main Idea	Details
Conditions for Life on Earth *I found this on page ________.*	**Restate** *the basic needs of Earth's life-forms.* 1. ________ 2. ________ 3. ________
I found this on page ________.	**Define** astrobiology. ________ ________
I found this on page ________.	**Identify** *three ways that Earth's atmosphere protects the planet's life-forms.*

1	2	3

Main Idea	Details
I found this on page ________.	**Relate** *the importance of water to life.* ________ water → ________ and ________ in cells
I found this on page ________.	**Describe** *how Earth's atmosphere enables water to exist as a liquid.* ________ ________ ________

Main Idea	Details

I found this on page ______.

Evaluate *conditions that most interest astrobiologists.*

Liquid water	Reason: •
Plentiful carbon	Reasons: • • forms ______ compounds in all living organisms

Looking for Life Elsewhere

I found this on page ______.

Explain *a scientific hypothesis about the possibility of liquid water on Mars.*

Observation	Hypothesis
Channels on Mars's surface look like streambeds.	

I found this on page ______.

Distinguish *two observations of potential evidence of liquid water on satellites in the solar system.*

Europa	Enceladus

Analyze It Summarize how Earth differs from other objects in the solar system in ways that allow it to support abundant life.

Lesson 3 Human Space Travel

Predict *three facts that will be discussed in Lesson 3 after reading the headings. Record your predictions in your Science Journal.*

Main Idea	Details

Technology and Early Space Travel

I found this on page ______.

Describe *technologies associated with space travel. Then, explain why* rockets *are used to launch* artificial satellites.

Artificial Satellite	Rocket

Explanation: ______________________

____

____

Robotic Space Probes

I found this on page ______.

Contrast *3 types of* space probes.

Space Probe		
Definition:		
Type: Description:	Type: Description:	Type: Description:

Main Idea	Details
Challenges for Humans in Space *I found this on page* ______.	**Evaluate** *the challenge of solar radiation for astronauts.*

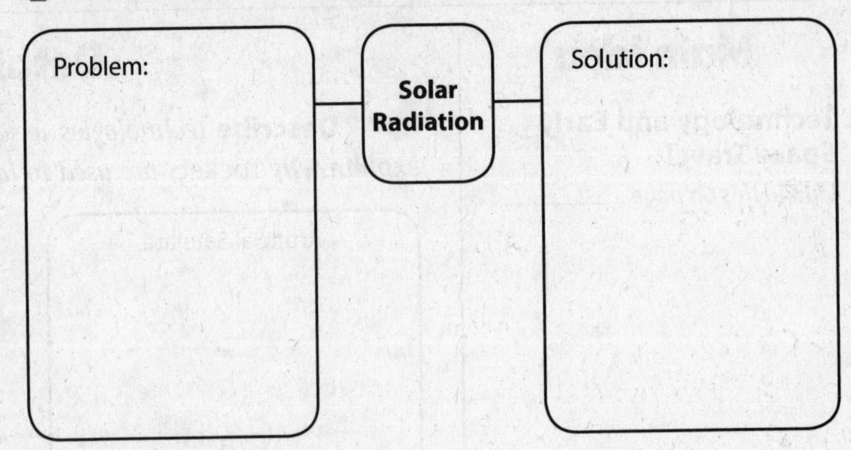

I found this on page ______.

Distinguish *ways human space travelers get the oxygen they need.*

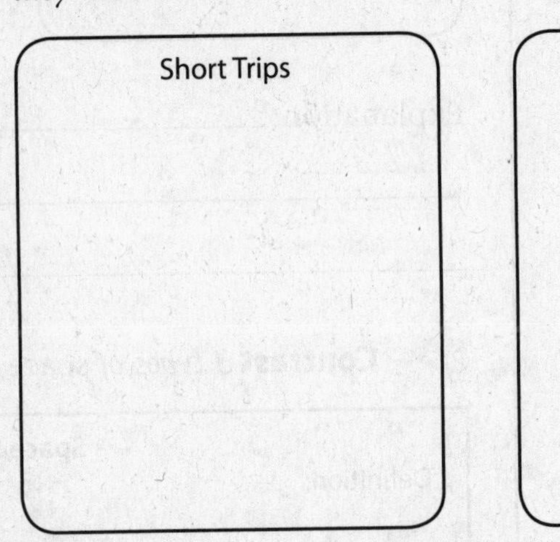

Longer Trips

I found this on page ______.

Identify *6 ways in which an astronaut's EMU suit provides protection.*

Extravehicular Mobility Unit
1.
2.
3.
4.
5.
6.

Main Idea	Details
I found this on page ______.	**Point out** *an advantage and two disadvantages of working in microgravity.* Advantage: ______ Disadvantages Short-term: ______ ______ Long-term: ______
Living and Working in Space I found this on page ______.	**Characterize** *the International Space Station (ISS).* ISS Use: Power source: Possible future use: Interior size: Crew size: Location:
I found this on page ______.	**Compare and contrast** *the space shuttle with the transportation systems used to send people to the Moon.* ______ ______ ______

Synthesize It Differentiate the challenges of uncrewed versus crewed space flight.

______

______

______

______

Review

Exploring Our Solar System

Chapter Wrap-Up

Now that you have read the chapter, think about what you have learned.

Use this checklist to help you study.

- ❑ Complete your Foldables® Chapter Project.
- ❑ Study your *Science Notebook* on this chapter.
- ❑ Study the definitions of vocabulary words.
- ❑ Reread the chapter, and review the charts, graphs, and illustrations.
- ❑ Review the Understanding Key Concepts at the end of each lesson.
- ❑ Look over the Chapter Review at the end of the chapter.

THE BIG IDEA

Summarize It Reread the chapter Big Idea and the lesson Key Concepts. Write a summary description of objects, life, and space travel within the solar system.

Challenge *Perform research to learn more about plans for commercial space travel of the future. Make a magazine-style advertisement of a space trip based on what you learn about the plans of real companies.*